SCIENTIFIC AND RELIGIOUS BELIEF

PHILOSOPHICAL STUDIES SERIES

Founded by Wilfrid S. Sellars and Keith Lehrer

VOLUME 59

SCIENTIFIC AND RELIGIOUS BELIEF

Edited by

PAUL WEINGARTNER

Institut für Wissenschaftstheorie,
International Research Center Salzburg

and

Department of Philosophy,
University of Salzburg

in collaboration with

Elena Klevakina-Uljanov

and

Gerhard Schurz

SPRINGER SCIENCE+BUSINESS MEDIA, B.V.

A C.I.P. Catalogue record for this book is available from the Library of Congress.

ISBN 978-0-7923-2595-6 ISBN 978-94-011-0804-1 (eBook)
DOI 10.1007/978-94-011-0804-1

Printed on acid-free paper

TABLE OF CONTENTS

PREFACE

The papers contained in this volume were given at the International Conference on Scientific and Religious Belief which was held at the Institut für Wissenschaftstheorie of the International Research Center Salzburg in August 1991. The conference was understood as a research conference dealing with an area which is not yet investigated sufficiently: Aspects of interrelation between scientific and religious belief.

The conference started with papers on important features of belief in general and on distinctive properties between belief, knowledge and acceptance. In the paper of Keith Lehrer they are analyzed comparatively and are also considered in respect to religious belief (faith). The paper of Franz von Kutschera takes up other important features of beliefs in general to be applicable to both, scientific and religious belief: reasons (explanations for beliefs which consist essentially in other beliefs again and which allow that the agent had alternatives) and causes (chance events or actions).

Gerhard Schurz's special topics on non-monotonic reasoning deals with the question how our beliefs change, when our information changes, especially when the respective information is increased or completely new information is added. Is belief value-free? Is value-ladenness an essential and intrinsic characteristic of belief? This question is dealt with in the paper by Elena Klevakina-Uljanov and applies to both scientific and religious belief.

The paper of Paul Weingartner tries to give an analysis of similarities and differences of scientific and religious belief. The analysis shows that this field of comparison is full of further interesting details with lots of new problems and questions not taken up frequently in recent literature. Joseph Maria Bocheński's paper deals with the important question of rational reasons and justification for religious hypotheses. One subquestion here is in what sense "religious experience" – if it exists – can be accepted as a justifying reason. This question is then continued in a detailed way by John Hick whose paper deals explicitly with the different features and aspects of religious experience not only in Christianity but also in the other great religions of mankind.

Paul Weingartner

ACKNOWLEDGEMENTS

My thanks go first to the participants for their kind cooperation and willingness to correct also the prepared extracts of the discussions. All their contributions are original except the one of Elena Klevakina-Uljanov which is an improved reprint from the volume *Advances in Scientific Philosophy*, edited by Gerhard Schurz and Georg J. W. Dorn. I take the opportunity to thank the publisher Editions Rodopi and Mr. Fred van der Zee for the kind permission.

Special thanks go to Elena Klevakina-Uljanov and Gerhard Schurz for their preparation of a part of the discussions.

Many thanks I want to say to the two secretaries: The difficult task of retyping the discussions from the tape was done by Sigrid Hinterstoisser and the final editorial corrections and the preparation of the camera-ready manuscript were realized by Gabriele Köbrunner-Krisch.

Last but not least I want to thank the International Research Center Salzburg for the financial support without which the conference could not have been organized.

Salzburg, July 1993 — Paul Weingartner

KEITH LEHRER

KNOWLEDGE, FAITH AND SKEPTICISM

Skepticism and knowledge stand in opposition as do faith and skepticism, and, consequently, both knowledge and faith as well as the relationship between them can be clarified by consideration of how they provide us with relief from skepticism. My contention is that the propositional attitude appropriate to knowledge is a different attitude than that involved in faith, though both attitudes are commonly referred to as belief. I distinguish between acceptance and belief, admitting that the distinction is to some extent stipulative, and argue that the former is the stuff of which knowledge is made. Faith, by contrast, involves belief and, according to some, the highest degree of belief, namely certainty.

The basic feature of acceptance is that it is a propositional attitude which is relativized to some goal or objective, the quest for truth, for example, in the strong sense that specific forms of acceptance are defined in terms of those definitions. One kind of acceptance, the kind germane to knowledge, just is, by definition, a positive propositional attitude and functional state aiming at obtaining truth and the avoiding error with respect to the proposition in question. Acceptance is the basis of the coherence theory of knowledge I defend.[1] The fact that faith and acceptance are different kinds of attitudes toward religious propositions means that there is no doxastic conversion of faith into knowledge. The attitudes are different, and one does not imply the other. I can accept what I do not believe, and I can refuse to accept what I believe, even what I believe in the highest degree. If, however, the attitudes appropriate to knowledge and faith are different, the objects of both may be the same. I may have faith that God exists and also know that God exists, though neither implies the other and each provides a relief from one form of skepticism.

The first section of the paper will contain my account, which contains a new attempt to characterize acceptance in terms of probability and functional role. I shall in this section characterize the relationship between acceptance and belief, the contrast between acceptance and faith. In the next section, I shall give an account of knowledge which is but a summary of my views on knowledge. In the final section, I shall argue that the propositional objects of faith may be the same. Let us turn first of all to an account of acceptance and faith.

P. Weingartner (ed.), Scientific and Religious Belief, 1–25.

ACCEPTANCE AND FAITH

Acceptance is the propositional attitude that is a condition of knowledge. Knowledge was traditionally analyzed as justified true belief and, since Gettier, some would simply supplement the analysis with some further condition. I turn to these matters below, but first I shall explain why belief should be replaced with a different propositional attitude in the analysis of knowledge. Knowledge implies truth, and the appropriate propositional attitude for knowledge should be one defined by that objective. Belief has many objectives. One might believe something, that one has a faithful spouse, for example, for reasons other than a concern for truth. A man might have very strong evidence of the unfaithfulness of his wife, having seen her drive off with a man looking into his eyes with the look of love and lust toward the hotel strip. But he might decide to believe in her fidelity nonetheless because it would make him too miserable to believe her unfaithful and because it would destroy the relationship. Or consider the example of religious faith. A person might believe in the existence of God while lacking what she considered to be adequate grounds for the truth of what she believes simply because it makes her happy to believe. Or, to take an example from Kierkegaard, she might believe that God exists with all her heart, and, therefore, be certain that God exists, not because of any objective evidence, she might think she had none, but because of a passion for her own eternal beatitude which is promised by her religious creed.[2] Of course, to believe something in these ways is, nevertheless, to believe it to be true, but, and this is critical, one may believe something to be true as the result of some motive other than a concern to obtain truth and avoid error. Hedonism or despair are often more than adequate motives for believing things to be true when one has no evidence of the truth of thing believed or, indeed, when one even has good evidence of the falsity of the thing in question.

Belief, because it may arise from indifference to or even in opposition to a concern for truth, is not what is required to attain knowledge. What is required is some attitude or state that by definition has truth and the avoidance of error as the objective. I have chosen the term "acceptance" to characterize this state. At times, I have thought of acceptance as a very special kind of belief, the kind which, by definition, has veracity has its objective, and if, for the sake of tradition, that is a comfortable way to think of the matter, I have no objection.[3] But acceptance, unlike many forms of belief, is, by definition, connected with truth seeking. One might, of course, speak of various forms of acceptance aiming at various objectives, but I am here restricting the use of the term "accept" to that sort of state that a person is in when he or she aims at truth.

More precisely, a person accepts that p in the requisite sense if and only if the person aims at accepting that p just in case p is true. The objective divides in two, to accept that p if p and to not accept that p if not p. We might, therefore, think of acceptance as the scientific or, perhaps, evidentially determined propositional attitude.

I should like to add some further clarification of the nature of this attitude and the relationship to belief. When I accept something in terms of the twin goals of obtaining truth and avoiding error, I evaluate and certify the proposition for a certain kind of functional role in thought, inference, and action. This evaluation and certification may arise from reflective judgement, but it can take place in a dependable but unreflective way as a result of processes that fall below the level of consciousness. Acceptance is not, however, the same thing as reflective judgement but is a functional state arising in some cases from such judgement. Acceptance is a functional state, a state that, like belief, has a specific function in thought and life. Reflective judgement is, by contrast, an episode in our intellectual life which may disappear without a trace when memory is diminished as in the aged, while acceptance a functional state that is maintained with assistance of memory to effect the appropriate result. Nevertheless, reflective judgement gives rise to acceptance in the normal intellect and is a familiar cause. Since the effect is less visible than the cause, it is useful to think of acceptance as the sort of state that arises from such judgement, though a state having the same function, the same influence, the same effect on thought, inference and action, may arise unreflectively in the course of receiving and evaluating information from various sources.

We may, however, characterize acceptance more precisely in terms of a probabilistic relationship to reflective judgement. The probability that one assigns to a proposition will, of course, influence how one thinks, infers and acts. If one assigns a high probability to some proposition, that will cause one to think of it as true, to infer further conclusions from it, and, in general, to act as thought it were true. The assignment of a high probability to a proposition is a necessary condition of acceptance, though it is not sufficient. The relationship between the high probability and acceptance may be explicated in terms of the stability of the probability assignment in the face of reflective judgements. Probability assignments may be evaluated, and evaluation may change them. Sometimes we assign probabilities, even quite high probabilities, on little evidence, or, in the extreme case on no evidence whatever. Acceptance is, however, a state that reflects evaluation and certification on the basis of our evidence and information. As a result, the high probabilities associated with acceptance already reflect the result of such evaluation and are stable with respect to such evaluations. This enables us to give a precise characterization of acceptance in terms of the stability of a high probability with respect to reflective judgement which explicitly and consciously incorporates evaluation and certification in terms of evidence and background information. We may say that a person accepts that p at t only if reflective judgement that p at t would not change the high probability assigned to p by the person at t. Belief, by contrast, may be unstable in that reflective judgement may cause an alteration in the assignment of the probability.

The foregoing characterization of acceptance brings out an interesting similarity between some forms of faith and acceptance. For some forms of faith provide a prophylactic against reflective judgement and are intended to do so. There is a form of faith that is simple and certain and unresponsive to reflective judgement. Thus, both simple faith and scientific acceptance have a similar stability. The difference, of course, is that scientific acceptance, or what I have simply called acceptance, results from the fact that all evidence and information has already be taken into account and, therefore, further reflection on the basis of that evidence and information will not, if the person is reasonable, change the probability she assigns. By contrast, the simple and certain faith, rather than taking such evidence and information into account provides an insulation against the effectiveness of it. Moreover, the stability associated with acceptance is synchronic only, it holds at a point in time, rather than having a dynamic effect over time. Of course, acceptance, and with it the assignment of probability, is sensitive to changes in evidence and information over time. The simple certainty of faith may, by contrast, be dynamically stable over time being invulnerable to the effect of new evidence and information.

We may summarize our account of acceptance by saying that acceptance that p at t incorporates a stable high probability for p at t resulting from evaluation and certification of p based on the background information at t of the person in question. More simply put, acceptance is evidential certification. Acceptance is, therefore, a propositional attitude involving higher order evaluation in terms of background information. This may occur quite unreflectively. I have suggested elsewhere that higher order evaluation may be thought of as the product of a central system that may evaluate propositions in terms of rules of thumb in a manner that, though appearing automatic, may be altered by ratiocination and the acquisition of new information.[4] I think of the matter this way. There are many doxastic states and propositional attitudes that arise naturally and become habitual. We may call these beliefs. They pass before the tribunal of truth. They are evaluated and when found wanting rejected. When the evaluation is positive, they are certified by the process and accepted. The natural states pass before the review of reason or the central system and are evaluated in terms of evidence and information to obtain truth and avoid error. Acceptance results from positive evaluation and becomes the basis of knowledge.

KNOWLEDGE AND SKEPTICISM

The account of knowledge I shall now offer is a summary of what I have developed elsewhere.[5] To abbreviate the exposition, I refer the reader to precise definitions listed in an appendix rather than repeating them here. Let us assume that a person has an acceptance system at a point in time. The acceptance system is a system of acceptances, acceptance that p, acceptance that q, and so forth. (D1.) The question is what we must add to acceptance that p at to yield knowledge that p at t, and the answer is a traditional one based on a coherence theory of justification. The basic idea is that justification arises when p coheres with some system, first of all, the acceptance system of the person. (D2.) Coherence is defined in a way that employs the skeptic as a dialectical foil. Theories of knowledge have always involved a confrontation with the skeptic, and, in my opinion, rightly so. My suggestion is that a person is justified in accepting that p, that is, p coheres with the acceptance system of a person just in case the objections of the skeptic can be met in terms of the acceptance system of the person. Justification and coherence thus reduce to meeting the objections of the skeptic.

If I accept something, that I see a computer, for example, which I now do, we may think of the skeptic as bringing forth objections which compete with what I accept, for example, that I am hallucinating, which would, if true, make it less reasonable for me to accept what I do. (D4.) One way to meet the objection of the skeptic is to appeal to the acceptance system one has. Taking a notion of one thing being more reasonable for me to accept than another on the basis of my acceptance system as a primitive notion, I can say that the competitor is beaten on the basis of my acceptance system. (D5.) The skeptical competitor, that I am hallucinating, is less reasonable for me to accept on the basis of my acceptance system than that I see a computer, primarily because I accept that I have not taken anything to make me hallucinate, that I can tell whether I am now hallucinating or not, that I am trustworthy in such evaluations and so forth. The crux of the matter is that acceptance incorporates evaluation in terms of my acceptance system which contains my evidence and information. The evaluation must, therefore, be based on a system that can meet the skeptical objections.

Beating a competitor amounts to the fact that it is more reasonable for me to accept the thing I do than what the skeptic objects on the basis of my acceptance system. That is a sufficient condition for meeting the skeptical objection on the basis of that system. It is not necessary, however. A subtle skeptic can raise an objection that it is reasonable enough to accept but which is not really germane. Suppose, for example, that skeptic objects that people sometimes dream that they see computers or even, what is true, of me that I sometimes dream that I see computers. I can meet this objection too but not by beating it, for I am sure that what the skeptic says is true. It is equally reasonable for me to believe that I see a computer and that I sometimes dream that I see computers, for I am sure that both are true. But though I cannot beat the objection, I can neutralize it with the reply that though I sometimes dream that I see computers, I am not dreaming now, and, indeed, it is just as reasonable for me to accept that I sometimes dream that I see computers but am not now dreaming as it is to accept the skeptics claim that I sometimes dream these things. I have neutralized the skeptics objection by rebutting his innuendo that I am now dreaming that I see a computer with the reply that I am not. Coherence with an acceptance system and, therefore, justification, or better put, personal justification in terms of my acceptance system, results when all the competitors, all the skeptical objections can be beaten or neutralized in terms of my acceptance systems. (D7.)

Thus, we have starting with acceptance, with the acceptance system of person, constructed a notion of personal justification. We cannot, however, equate knowledge with personally justified true acceptance. The reason, of course, is that it is possible that the acceptance which generates replies to the skeptical objections might be an acceptance system of things that are erroneous. So we must add that even when the acceptance system of a person is altered either in part or in total either by eliminating acceptance of error or replacing acceptance of error with the acceptance of truth, the resulting system or systems, which I call members the ultrasystem of the person, still suffice to beat and neutralize all competitors. (D10.) When all such competitors remain beaten or neutralized, I say that the personal justification is undefeated. (D8). When not, then I say that the personal justification is defeated. (D9.) Knowledge, I propose, is undefeated justification of the acceptance of a truth. This proposal reduces, however, to a simpler one. For undefeated justification implies both acceptance and truth, the former because acceptance is required for justification and the latter because error would result in defeat. Hence, from an extensive analysis we obtain the elegantly simple result that knowledge is undefeated justification.

Before applying these results to the issue of faith, it is necessary to consider some objections. It is has been objected that higher order evaluation will lead to a regress of evaluation upon evaluation, and it has been argued that coherence among acceptance might be isolated from the world.[6] The replies to both are connected. One competitor for any claim whether about the inner world of consciousness, the outer world of perception, the unperceived world of theory, or even the supernatural world, is the claim that one is not competent, not trustworthy about such matters, perhaps because one is isolated from the reality in question if such exists. Such a competitor must be beaten on the account I offer for personal justification to arise, and that personal justification must be retained in the ultrasystem. The retention in the ultrasystem requires the truth of the claim that one is justified which is assumed to beat the competitor. Even the neutralization of the weaker competitor that we are sometimes untrustworthy in what we accept requires the acceptance of the claim that one is trustworthy in the matter under consideration. Simply put, then, if I accept that I think, I must accept and be right in accepting that I am trustworthy in what I thus accept, if I accept that I see a computer, I must accept and be correct in accepting that I am trustworthy in what I thus accept, if I accept that neutrinos exist, I must accept and be correct in accepting that I am trustworthy in what I thus accept, and, finally, if I accept that God exists, then I must accept and be correct in accepting that I am trustworthy in what I thus accept. If I do not accept that I am trustworthy, then the skeptical objection that I am untrustworthy will not be met and I will not be personally justified. If I am not correct in accepting that I am trustworthy, then my personal justification will be defeated. Thus, the isolation objection gets met when personal justification gets transformed into undefeated justification, for this requires the truth of the claim that I am trustworthy which insures that I not isolated from the reality I accept. In short, all knowledge requires that I am trustworthy in what I accept.

The foregoing remarks provide the basis for meeting the regress objection. The objection might be formulated as follows. You suppose that you are trustworthy in accepting that p. You admit that if you are not then you are not justified in accepting that p. But are you trustworthy in accepting that you are trustworthy? If not, then your acceptance of your trustworthiness would itself be unsupported, and being unsupported can hardly support the acceptance of p. By way of reply, consider the principle

T. I am trustworthy in what I accept.

We may understand this principle as affirming that if I accept that p, then p is more worthy of my acceptance than the denial of p. Suppose that I accept T. Am I trustworthy in what I accept? What I accept, which includes T, tells me that I am, that T is more worthy of my acceptance than the denial of T. Hence if T is part of my acceptance system, it suffices to end a regress by vouching for itself. Principle T tells me that I am trustworthy in my acceptance of T because that is one of the things that I accept. Hence principle T has the important feature that acceptance of it supports itself. Moreover, if I accept T, and T is true, and, finally, if I accept that it is true because it is true, that is, if I accept that I am trustworthy in what I accept because of my trustworthiness in what I accept, the way appears clear for the conclusion that I know that principle T is true.

It might be doubted that people do accept principle T. It may seem like a very intellectual principle and beyond the intellectual ken of most. But that would be a mistake. The fact that people do trust what they accept and do trust themselves in what they accept, shows that they do accept that they and what they accept are worthy of their trust. They might not readily come up with a formation of principle T, but they are in a functional state of accepting that p. For that functional state just is the state that produces the results in thought, inference and action, resulting form their trust in themselves. It might also be thought that principle T is too unrestricted. That we think that we are trustworthy in some of things we accept, or some kinds of things we accept, but not others. That objection rests on a misunderstanding of acceptance. Certification based on evaluation in terms of evidence and background information aiming at the avoidance error, at not accepting that p if p is false, already precludes acceptance that is untrustworthy. In short, acceptance based on critical evaluation is certification restricted to cases in which the evidence and information puts one in a position to accept what one does in a trustworthy manner. Thus, principle T suffices to meet the two objections, the isolation argument and the regress argument, which are the major objections to the coherence theory of justification and knowledge. With a theory of knowledge and acceptance before us, let us return to the matter of knowledge and faith.

KNOWLEDGE AND FAITH

What is the relationship between knowledge and faith? The attitudes of faith, belief and even certainty, are the wrong sort of attitudes to yield knowledge. Faith and even certainty that God exists do not convert into knowledge when they arise from motives other than obtaining truth and avoiding error, in short, veracity. Security and piety are often the motives of faith, and they are different from motive of veracity. But one should not conclude from this that faith is incompatible with knowledge, that faith and certainty are incompatible with acceptance and justification. Some theists have placed great emphasis on the irrationality of faith and the irrelevance of reason and objectivity to faith. Others have been more sympathetic to uses of reason. But the sort of account I have articulated above suggests a compatibility between religious belief and scientific acceptance. Belief may have piety as a motive while acceptance is motivated by veracity. But the very same proposition, that God exists, may be both believed out piety and accepted out of veracity. The attitudes of belief and acceptance are different, but the object of both may, as I noted above, be the same. Of course, concern with acceptance may interfere with the certainty of faith. That is a question of some psychological complexity and controversy, but the logic is clear. There is no logical incompatibility between religious belief and scientific acceptance. My own experience suggests to me that psychological conflict between belief and acceptance is also exaggerated in my opinion. Belief supported by acceptance may be less heroic belief than belief that stands in conflict with acceptance, but who is to say that acceptance, that the rational acceptance of the existence of God, may not for some, perhaps for many, render the step of piety and faith steadier and more natural? And who is to say that belief might not lead one to consider evidence and information that would lead one to acceptance? For my part, I am opposed to those who would oppose belief and acceptance, veracity and piety to each other and pit one against the other. Suppose that God exists and is responsible for the construction of our mind and body. Are we supposed to believe that he gave us one attitude, acceptance, aimed at veracity, and another attitude, belief, aimed at piety, and placed powers within us to pit one against the other? Surely, our maker, if such there be, would be kinder and revealing his existence to us would provide us with the evidence so that belief and acceptance, faith and reason would combine to the same conclusion. Or so it seems to me.

What about knowledge? Suppose that faith, perhaps based on the feelings of divine presence, has given a person certainty that God exists. Does she know that God exists? She will say so, for when people are certain of something, they say they know. But often they do not. Certainty is not the same thing as knowledge, though it may feel like it is. Some philosophers are inclined to defend the claim that the woman knows if her belief arises in a reliable way or is the product of some natural faculty. I think that is incorrect. The person may have no idea that the idea is reliable, has arisen in a reliable manner or is the product of some natural faculty. These may supply her with correct information, but suppose she has no idea that this information correct. She believes, of course, but, let us suppose, she does not know whether the information she has is correct. She has received the information that God exists but does not know that the information is correct. Then she does not know that God exists, though she may be certain enough. What is required for her to know? It must be more reasonable for her to accept that her information is correct than that it is untrustworthy on the basis of her evidence and information. It must be more reasonable to accept that she is trustworthy in the matter than that she is not. For if it is no more reasonable for her to accept the conclusion of atheism than the conclusion of theism on her evidence and information, then she is ignorant, however pious she may be. Suppose that she accepts her trustworthiness in what she accepts on the basis of her evidence and information, and, moreover, suppose that she is correct in accepting this. She is trustworthy as she accepts and trustworthy in the way that she accepts. Then her personal justification for belief in the existence will be undefeated, and she will have knowledge.

It is perhaps worth noting that a person without faith might also know that God exists. She might reflect on argumentation, perhaps undergo some religious experience, but she might fear the existence of God. Fear keeps her from faith and belief. But evaluation of the conclusion that God exists on the basis of her evidence and information leads her to certify that conclusion in the interests of veracity. She thinks and reasons with the pious, but she withholds belief. This is the odder case, for, as I have said, acceptance seems like a kind of belief. Yet I think we can easily understand how this second woman fails to arrive at the attitude toward the conclusion that God exists that pious refer to as belief and faith. The point is that she might never believe in the existence of God out of piety, and, in this case, her acceptance of the existence of God out of a concern for veracity contrasts sharply with the belief of the pious.

I have drawn the conclusion that knowledge and faith combine rather than stand in opposition. There is a coherence within acceptance that yields knowledge, and such acceptance is systematic. The acceptance system of the theist differs from the acceptance system of the atheist in more than the simple claim that God exists. Each has a view about how one can tell, about when one is trustworthy and when not in such matters, and in this they differ. The acceptance system of each differs at a metalevel as well as at the first level. The one evaluates propositions differently from the other and rightly so. In this matter as in other matters, mental ascendance beyond the first level is decisive. Religion is a special form of metamental transcendence. So is the opposite.

NOTES

1 K. Lehrer ([1]1990).
2 S. Kierkegaard (1946), 220.
3 K. Lehrer ([1]1990), Chapter 2.
4 K. Lehrer, ([2]1990).
5 K. Lehrer ([1]1990), Chapters, 6, 7 and 8 and Lehrer ([2]1990), Chapter 11.
6 Cf. P. K. Moser (1989) and G. J. Mattey (1989). The objection originates from J. Pollock (1974), 27-28, but is rejected in later work (Pollock, 1986).

REFERENCES

Kierkegaard, S.: 1946, *A Kierkegaard Anthology*, R. Bretall (ed.), Westview Press, Princeton, N.J., 125-138.
Lehrer, K.: [1]1990, *Theory of Knowledge*, Routledge, Boulder and London.
Lehrer, K.: [2]1990, *Metamind*, Clarendon Press, Oxford.
Moser, P. K.: 1989, 'Lehrer's Coherentism and the Isolation Objection', in *The Current State of the Coherence Theory: Critical Essays on the Epistemic Theories of Keith Lehrer and Laurence Bonjour, with Replies*, J. W. Bender (ed.), Reidel, Dordrecht and Boston.
Mattey, G. J: 1989, 'Personal Coherence, Objectivity and Reliability', in *The Current State of the Coherence Theory: Critical Essays on the Epistemic Theories of Keith Lehrer and Laurence Bonjour, with Replies*, J. W. Bender (ed.), Reidel, Dordrecht and Boston.
Pollock, J.: 1974, *Knowledge and Justification*, Princeton University Press, Princeton, N.J.
Pollock, J.: 1986, *Contemporary Theories of Knowledge*, Rowman and Littlefield, Totowa, N.J.

APPENDIX: DEFINITIONS

D1. A system X is an acceptance system of S if and only if X contains just statements of the form – S accepts that p – attributing to S just those things that S accepts with the objective of accepting that p if and only if p.
D2. S is justified in accepting p at t on the basis of system X of S at t if and only if p coheres with X of S at t.
D3. S is justified in accepting p at t on the basis of system X of S at t if and only if all competitors of p are beaten or neutralized for S on X at t.
D4. c competes with p for S on X at t if and only if it is more reasonable for S to accept that p on the assumption that c is false than on the assumption that c is true, on the basis of X at t.
D5. p beats c for S on X at t if and only if c competes with p for S on X at t, and it is more reasonable for S to accept p than to accept c on X at t.
D6. n neutralizes c as a competitor of p for S on X at t if and only if c competes with p for S on X at t, the conjunction of c and n does not compete with p for S on X at t, and it is as reasonable for S to accept the conjunction of c and n as to accept c alone on X at t.
D7. S is personally justified in accepting that p at t if and only if S is justified in accepting that p on the basis of the acceptance system of S at t.
D8. S is justified in accepting that p at t in a way that is undefeated if and only if S is justified in accepting p at t on the basis of every system that is a member of the ultrasystem of S at t.
D9. M defeats the personal justification of S for accepting p at t if and only if S is personally justified in accepting p at t, but S is not justified in accepting p at t on system M at t where M is member of the ultrasystem of S at t.
D10. A system M is a member of the ultrasystem of S at t if and only if either M is the acceptance system of S at t or results from eliminating one or more statements of the form – S accepts that q – when q is false, replacing one or more statements of the form – S accepts that q – with a statement of the form – S accepts that not-q – when q is false, or any combination of such eliminations and replacements in the acceptance system of S at t with the constraint that if q logically entails r which is false and also accepted, then – S accepts that r – must also be eliminated or replaced just as – S accepts that q – was.

These definitions yield the following definition of knowledge:
DK. S knows that p at t if and only if (i) S accepts that p, (ii) it is true that p, (iii) S is personally justified in accepting that p at t and (iv) S is justified in accepting that p at t in a way that is undefeated.
Condition (iv) implies condition (iii), and interpreting it so that it implies (i), it will imply (ii) as well. So, we may obtain the following reduction of the analysis:
DK*. S knows that p at t if and only if S is justified in accepting that p at t in a way that is undefeated.

DISCUSSION

HICK: When you say that my acceptance of *p* is an undefeated acceptance, this entails that *p* is true – did you want to say that?

LEHRER: Not exactly. What I wanted to say was 'the undefeated justification' rather than 'the undefeated acceptance.'

HICK: Are you not assuming that undefeated equals undefeatable – that is to say, an acceptance may be undefeated in fact, but there might be some defeater of which I am not aware?

LEHRER: There might be some, of which I am not aware.

HICK: But then one is not entitled to say that undefeated justifications are necessarily true.

LEHRER: It may be the case that you are not aware of some defeater, but it is nevertheless a defeater, even if you are not aware of it.

So perhaps the best way to put it is that at a given point in time, from my first person perspective, all you have access to is information about personal justification.

So it can be that your justification is defeated even though you do not recognize that and are not in a position to recognize it.

HICK: But are you in that case entitled to add the condition of truth to your definition of knowledge, when you define knowledge as undefeated justification?

LEHRER: If I accept *p* and *p* is in fact false – I might be personally justified in accepting *p*, but the falsity of *p* is in fact the defeater, even though I do not recognize it to be so.

HICK: But surely your definition is undefeated, meaning undefeated in fact, not meaning undefeatable in principle.

LEHRER: Undefeated in fact means undefeated in terms of what really is true or false – not just in terms of what I think to be true or false. So, if in fact I am untrustworthy, although I take myself to be trustworthy, then the skeptic wins.

HICK: But may you not be entirely trustworthy in terms of the information you have, and yet may there not be further information that you do not have, and do not know that you do not have, such that your undefeated justification does not equal truth?

LEHRER: I do not think it can be. Suppose I accept *p* and *p* is false. Then there is a correction in my acceptance system, where you replace or substitute 'I accept that not *p*' for 'I accept that *p*.'

HICK: But this is only if it comes to light, that *p* is false.

LEHRER: The condition is really an external constraint, it is not something that I am always in position to ascertain.

HICK: Surely not, because there might be a defeater that has not come over your intellectual horizon, so that you do not know of it at all.

LEHRER: But as I am using 'defeat' in that case, I would say, it *is* defeated, I just do not know *that* it is.

HICK: So, you are trustworthy given the information you have and your justification is undefeated. But, nevertheless, your conclusion is false.

LEHRER: No. If the justification is undefeated, the conclusion is true. There is some sort of misunderstanding, but I have not succeeded in clearing it up. Let us stick with your formulation of it. What conclusion would you draw?

HICK: My conclusion would be that one must not include in knowledge-conditions that what you know is true. For you are using 'knowledge' in such a sense that what you know, under your definition, may not be true.

LEHRER: What I mean by undefeated justification is what you mean by undefeatable justification.

HICK: But then we never know that we have it.

LEHRER: So, it might turn out on this account that we *do* know, but your point is that we would never then know *that* we know.

HICK: Something like that, yes. Is this the fallibilism of which you were speaking at the end?

LEHRER: I think that *is* a kind of fallibilism. The notion of knowledge that I put forth could be applied to itself, that is, it might be in fact the case that there is not anything that would defeat my justification. Now, suppose I also hold the view that it is so and, moreover, suppose my reasons for that are really perfectly good reasons. In that case I might not only know, but might also know that I know. Maybe I should put it this way: It is a really revisionist notion of knowledge. Because when we talk about knowing that we know in an ordinary language there is the implication that it could not be wrong. But in this respect I am a skeptic.

And since this is a fallibilist notion – although it is the case that it might be possible to know and also to know that we know in my sense – we nevertheless could have been wrong, in a certain sense.

HICK: There is the view that 'A knows *p*' entails that *p* is true. But that is an ideal, and we can never know that we attain the ideal. And therefore it is a mistake to include the truth of what we believe within the concept of knowledge. For then we should always have to say, 'I claim to know.' For we have no access to the truth other than whatever it is that justifies our beliefs. So we can never get ourselves out of the circle of having justified beliefs that may be true but on the other hand may not be.

KUTSCHERA: I do not see where your definition is very different from the one of knowledge as true belief. You made a distinction between belief and acceptance. But one would usually say that if I believe that *p*, I believe that *p* is true. So, belief has something to do with truth. And the question of whether I automatically believe something or not, does not seem very important to me in this respect. Except from this distinction then we have it that the subject of knowledge has to believe that *p* and it has to be true, i.e., in your terminology 'undefeated'. Personal justification does not add anything to belief – at least if belief is taken in an ideal, rationalistic way, so that we cannot believe what is inconsistent etc.; if, e.g., we take belief in a way explicated in standard doxastic logic. Then we would also have something like a principle of trustworthiness: I believe that what I believe is true. That is a principle of doxastic logic. Justified belief adds to true belief only if we refer to intersubjective standards of justified belief, standards for rationally forming beliefs. The best justification, of course, would be truth, but you would not add anything to the second condition for knowledge, namely truth, by that. So, if you just speak of personal justification, I do not see, that you will get very far: I am personally justified in believing *p* if *p* is true according to the lights of my beliefs.

LEHRER: Let me start where you were discussing the relationship between belief and acceptance. I do not want to disagree greatly with what you said there. I do think that there has been a notion of belief that has played a quite significant role in epistemic and doxastic logic. And I agree with you that *believing that p* and *believing that p is true* are very close. Still there is a peculiarity: You can believe that *p* is true without that being your motive. There can be cases like in the example of the husband who believes that his wife is faithful. Because he also believes that it is true that his wife is faithful. But the very reasons for that belief, the basis of that belief is not a concern for truth, it is a concern for his felicity. He would be very troubled if he thinks that his reasons for believing that his wife is faithful and for believing that it is true that his wife is faithful are not reasons that point to the truth. So, I think that the objective or the goal of the doxastic state is very important. I do not think that that is described adequately by the observation that believing something and believing that it is true is the same thing. So, I think you really need to focus on the goal, the truth seeking goal.

Now, with respect to this proposition, '*a* believes that if *a* believes that *p*, then *p*', I think it is a bit strong, if one is a fallibilist, for the following reason: If you would generalize that, it would seem to me, that it would be false, i.e.: It is not the case that I believe that everything that I believe is true. In fact, on the contrary, I am quite sure, that some of the things that I believe are false. So, I think the notion of being trustworthy is a little bit weaker than the notion that says that we only believe something that is true.

KUTSCHERA: If I believe that *p* is true I can still believe, that *p* is possibly false, i.e. contingent.

LEHRER: But I just believe that some of the things I believe are possibly false. If you look at your belief system historically then it does seem very obvious that it is not going to be perfect. So, I will be amazed that everything what I accept is true. I do not think that everything that I accept as such is worthy of my acceptance. What gives me the justification is not just the accepting of *p*. I might not have the kind of support that is necessary for knowledge.

KUTSCHERA: $a\mathrm{B}\ (\forall p\ (a\mathrm{B}p \supset p))$ is not a consequence of the principle: $a\mathrm{B}\ (a\mathrm{B}p \supset p)$ for each *p* in standard doxastic logic.

LEHRER: I think the first order system is inadequate. I think what is peculiar about us is not just the way we believe and accept things, but that we evaluate our acceptance and belief of them and we evaluate the terms of the background system, which in effect contains information about when we are likely to go wrong and when we are not likely to go wrong. Justification results from a systematic influence that goes beyond the mere acceptance of *p* and the other first order acceptances to higher order evaluations. That is why I have raised the point about trustworthiness. Because, if you have just this full first order system but no evaluation of it, then, of course, there is nothing to protect you from the claim that the whole set of first-order beliefs is true just because of sheer luck, the claim that it is just one big guess.

KUTSCHERA: It is the same as with money. If you have earned a lot of money people might say that you are just lucky. You yourself would not see it as sheer luck mostly. You would say, I have put a lot of diligent work into getting it. We also usually put a lot of work into your epistemic systems and so we might say: Well, it is not luck at all, if we are right. Trustworthiness is only subjective for me. I trust myself, I trust my beliefs to be right, as we have seen, but I am not thereby trustworthy for other persons. A supposition I do not trust in is not a belief in the strong sense of a conviction.

LEHRER: Your claim about your trustworthiness is a claim that can be right or wrong, like the claim that there is a table in front of me can be right or wrong. If it is right, then it will yield the kind of justification we required. If it is wrong, then we are out of luck.

BOCHENSKI: Just to clarify the meaning of the word we are using: There is a distinction between belief and acceptance. As far as I understood you correctly, acceptance is connected more closely with truth than belief. You cannot believe something without believing that it is true.

LEHRER: If you believe something, then you believe that it is true. But I do not think that truth has to be your motive of belief. So I take the case described by Kierkegaard. He says that Abraham goes up the mountain, believing that he is going to sacrifice Isaac, but also believing that Isaac is going to live. Now he says that this is absurd. This is an extreme case and that is why I take this other case, where the man believes that his wife is faithful. That makes him happy and improves the relationship. But, of course, nobody believes that he believes out of an interest in what is true. It is just that it is not an interest in truth, that motivates him. So, I am just using acceptance as a kind of belief-like state resulting from an interest in truth. We are motivated to get the truth. You accept something to get at the truth.

BOCHENSKI: You do not think that you can accept a propositon without any relation to truth. If I may use a personal flight experience. Flying to a pass, I found it closed. But I knew there is another pass 50 kilometers to the left, but did not know anything about the truth of the proposition saying that it is open. What would you say to that? I did accept the said proposition because I had to act.

LEHRER: I think it is not the notion of acceptance that I want to utilize, because the problem here is that your alternatives are limited and you look at your alternatives and you have to ask yourself: Well, which alternative maximizes the expected value for me, which, in this case, is, which alternative gives me the better chances of surviving? And *that* alternative will give me the better chances of surviving, but still they may not be very good.

BOCHENSKI: Yes, that is another possibility. Anyway, my opinion is that belief is somewhat more strongly linked to truth than is acceptance. I have the feeling that when somebody says, I believe, another one says, I accept, the relation to truth is a closer in the first case.

LEHRER: There is a difference between first and third person usages here and I am inclined to be a little mistrustful of our first person usages, because they are connected with speech acts and speech act implications in a way that is misleading with respect to semantics. When I say, I believe, then I suspect that is a stronger claim. That is right.

SCHURZ: I want to go back to the point of Kutschera. I do not know whether I have got it right, but one message of his questions seems to be the following problem: Doesn't the whole system of your theory of knowledge collapse into the notion of true belief? Does the notion of personal justification add anything to the notion of true belief? The point involved in Kutschera's question seems to be a deep one. Being justified, in your definition, means for instance not to be beaten by a competitor, etc. And this definition goes back to a primitive notion. This primitive notion is "is more reasonable than": some proposition is more reasonable than another proposition. And Kutschera's point was, that the notion of justification has to involve something which is more than just personal belief. Otherwise the definition would just collapse into mere true belief. And I think that is really so, because one can prove a theorem about your definitions right ahead, namely: For every system of true belief there is a personal probability function that makes it justified in your sense. The fact that I believe p and not q just implies that p cannot be less reasonable than q – if reasonability is explicated in terms of subjective probability. If you add the condition that p is true and q is false, then everything is done for your system, and you have undefeated justification. Let us give an example: The weather has been fine all the days until today. Then one believes that it will be fine tomorrow – without any reason. For the one person it is more reasonable to believe that the sun will shine tomorrow, for the other person it is more reasonable to believe that it will rain tomorrow. So, if the sun will really shine tomorrow the one person has knowledge in your sense, if it will rain tomorrow, the other person has knowledge in your sense, but there no notion of justification which goes beyond the mere belief systems. So, my question is: What does prevent your definition of knowledge to collapse into a system of true belief?

LEHRER: What I have been eager to do is to try to get knowledge out of true belief plus a normative notion. Your are quite right, the normative notion is going to do some work in the theory. Whenever a person believes, we can imagine that belief supplemented with a belief system. If the belief is true, we can imagine that belief to be supplemented with the belief system such that, if that belief system turns out to be true, then the person knows that the belief is true. The question is: Is that a problem?

Why should it not be the case that if I have a belief system of a certain sort, then I know? I do not see any strong objection to that, provided that this belief system of mine is not restricted to first-order beliefs. It might be a belief-system that will tell us that, whenever I have beliefs of the sort that I do have in the circumstances that I am in, which, moreover, I can recognize, then I am virtually always right about the matter. I might have a view not only about *p*, but about my cognitive relationship to propositions like *p*. So, with respect to the case of the weather, if I have a belief system, and the belief system is wrong, that says that under certain circumstances I can predict a very high degree of accuracy that I am correct and tells how I do that. And that system is, moreover, correct. Then, it seems to me, I *do* know. So, what you are supposing is that this background system might be rather inadequate. But remember, every objection is going to be met. So, one objection, e.g., is the one that you cannot tell whether it is going to rain tomorrow or not. That is an objection's competitor. You believe that it is going to rain tomorrow. One objection to that is that you can not tell whether it is going to rain tomorrow or not. Whatever it is in your system that is to meet that objection, it has to be true. Otherwise justification of it is defeated. So, there is a trick in my theory: In order to meet the objections, in order to have personal justification, your system has got to be able to meet all the skeptical objections. In order to have the justification being undefeated, what meets those skeptical objections, it must be true. This is a real constraint on what that belief system has to be like. It has to be adequate to meet every objection and to be undefeated.

SCHURZ: I do not agree, because, in this case, what is an objection in your system depends on your preference evaluations of what is more reasonable. For me, it is not more reasonable to believe that it is raining tomorrow than just to believe that I cannot predict it. That is not an objection for your system, because it is not true in your system. You must only meet all *your* objections. But you can just choose your cognitive preference relation, *p* is more reasonable than *q*, in a way that all my objections are not objections for you. That is a consequence of the system, isn't it? And if this is a consequence, then your answer is not really satisfactory.

Kutschera's point was just this: You need a notion of cognitive preference or a justification which is just a little bit more than belief.

LEHRER: That is a good and deep objection. So the point that you are making is that certain things which I have intuitively assumed would constitute skeptical objections, might turn out not to constitute skeptical objections for a person. The only thing that I could say to that – it is not really adequate – is that in this justification that the person has will be certain errors. And when those errors are corrected then things will turn into competitors. If your are right then that person might turn out to be personally justified too deeply at too little expense of the person. But I then have to argue that the acceptance system would have to contain errors, so that, when they were corrected, the competition would arise, i.e., the intuitive objections would turn into objections and then would be all met.

(Added after discussion.) The reply to Schurz now seems clear to me. His claim is that what constitutes an objection or competitor to a claim on my account will depend on my acceptance system, and, more specifically, upon my accepting the relevance of the objection or competitor. He then proposes that if I do not accept the relevance of any competitors, then there will be no competitors to be beaten or neutralized. In that case an acceptance may slip over too easily into personal justification and undefeated justification just because it has so few competitors.

The objection is important for it brings out the central role of my acceptance of my own trustworthiness. It is my acceptance of my own trustworthiness that makes one thing more reasonable for me to accept than another, or so I have argued. If I do not accept that I am trustworthy in what I accept, even with some qualifications, then, no matter what I accept, nothing is any more reasonable than anything else for me to accept. Hence, if I do not accept my own trustworthiness regarding acceptance, then it is no more reasonable in terms of my acceptance system for me to accept that *p* than to accept the denial of *p*. In that case, my acceptance of *p* will not beat any competitors. Now Schurz may reply that my acceptance of *p* will not have any competitors in this case either, and an adequate reply requires a minor modification of my requirements for personal justification.

To meet the objection, I need to amend how I have defined personal justification in a minor way by adding the requirement that any proposition must have and beat one competitor, namely, its denial. The requirement is as follows: it is more reasonable on the basis of the acceptance system of S at *t* for S to accept that *p* than to accept the denial of *p*. Let us call that the minimal success condition. For that condition to be satisfied, S must accept that S is trustworthy in what S accepts. Moreover, for S to know that *p*, S must also be correct in accepting that he is trustworthy in what he accepts or else his personal justification for accepting that *p* will be defeated. So for the minimal success condition to be satisfied, S must accept that S is trustworthy in what he accepts and, to obtain knowledge, S must be correct in accepting this.

How does this offer a reply to Schurz? Might I not accept that I am trustworthy in what I accept and be correct in this even though I do not accept the relevance of objections to my claims? No! To be trustworthy in what one accepts requires that one accepts the relevance of genuinely relevant objections and meet those objections. The acceptance of my own trustworthiness is necessary for satisfaction of the minimal success condition, and the correctness of the acceptance of my trustworthiness will depend on my accepting the relevance of relevant objections. So, the objection can be met but, as I concede, only be modifying personal justification by adding the minimal success condition.

KLEVAKINA-ULJANOV: As you pointed out, truth is not the motive for a belief. The term 'belief' is a so-called epistemic term and is related to knowledge or truth as a goal. From this point of view, what would you say concerning belief? Maybe, in this way it is really related to truth as a goal and then truth is the motive for the active belief.

LEHRER: I certainly think that truth is often a motive for belief. But I also think that it is more important that fairly often belief seems to me just to be unmotivated, i.e., it just arises in a very natural, automatic, and unreflective way. So, for me it does not matter too much whether we should use the term 'acceptance' or stick with the notion of belief. But just add the qualification, that belief is motivated or arrived at by an interest in truth. That will preclude beliefs that arise automatically or beliefs that are arrived at for some interest or some purpose, even ones contrary to truth. There are such beliefs. Don't you think that if a person keeping saying something over and over, she might wind up believing it? Suppose she does. Then the belief may not have been motivated by truth. It may just have resulted from verbal repetition. But, suppose I observe this feature that if you keep saying something often enough you wind up believing it. And now there is something that I would like to believe, but I don't. So, I keep saying it and it works, I wind up believing it. Again, I may not have been motivated by an interest in truth. I just may have been motivated by wanting to believe.

KLEVAKINA-ULJANOV: Yes, I agree with you. Certainly, very often there are such cases that belief is completely unmotivated, or it is not motivated by truth-seeking goals. But is it possible to speak about belief and acceptance in general and to say that belief can be distinguished from acceptance, and that acceptance is truth-oriented and belief is not? There are too many variants of kinds of belief, and if you take scientific belief, is it possible to say that there is no relation to truth-seeking?

LEHRER: I do not think that I disagree with you. In the case in which belief is truth-seeking belief there is no interesting difference between acceptance, as I have talked about it, and belief.

I think the most significant thing I have to say about it actually was the relationship between probability and reflective judgement. There are lots of beliefs that we arrive at because somebody just tells us something. We do not think about what they tell us, and we just believe them. But if we would reflectively judge the matter, then that would create some doubts. So, reflective judgement in this case would change the probability which we would attach to the original judgement. That original judgement is a sort of unreflective and unscientific belief. When, on the contrary, I really believe scientifically, something evaluated scientifically, then reflective judgement would not change the probability at all. This is acceptance. So this perhaps covers some way of explicating a special notion of belief which I call acceptance.

HIEKE: You have to make a distinction between the conceptional contents of these terms, 'belief' and 'acceptance'. So, 'acceptance' is really a primitive term of your theory and then you have to have some principles which implicitly define your term. I refer to your book now – do you accept the principle 'If *b* accepts *p* then *b* believes *p*'?

LEHRER: Let us take that as a principle. So acceptance is a special kind of belief.

HIEKE: You made three points with which you implicitly defined acceptance: The first point was truth-orientation – the one you disputed about now – the second point was that it imposes a certain kind of rationality upon itself and the third point was that it stabilizes probability. This should be your three main principles concerning 'acceptance'. Now, my question is just one of clarification: Can you provide some little more formal versions of these principles?

LEHRER: I do not have any well worked out logical principles of acceptance. I was trying to get some clarity about an object that one might formalize. So, I could articulate those principles, i.e., the ones you mentioned with some precision. I thought about this notion of conscientious judgement which is closely related to that one of acceptance, but they are not the same thing, since we obviously accept many things without having reflected on them. But we can do it intelligently and scientifically without extended reflection. So, the point is that acceptance has the same functional role in action and inference. I would allow the formalization to be probability formalizations and expected epistemic utility theory. Just to give an example: I can prove to you in terms of probability relations that these background acceptance systems have a certain equivalence in terms of prior probabilities in the following way: Compare a skeptical hypothesis to a common-sense hypothesis about there being a table here. Most of our acceptance systems are such that they tell us that on the basis of our sensory evidence it is highly probable that there is a table here and not very highly probable that we are being manipulated by a demon. So, the common-sense hypothesis is more probable than the skeptical one. If you say that, then I think that I can show you that it is a function of your prior probability distribution. So, in order to get that probability distribution, you have to assign a higher prior probability to common-sense than to the demon. The point I am trying to make is that once you connect acceptance with probability, the rest of the formalization can be a formalization in terms of the manipulation of probability relations – how you conditionalized your experience and how you resolve conflicts. So, the answer is: I have got those principles, I could give you a formal account of them.

BRANDL: I have some short comments on three points. The first one is mostly terminological. According to your definition, the acceptance system of *s* is not a set of propositions that *s* accepts, but a set of propositions of the form '*s* accepts that *p*'. It is therefore strictly speaking incorrect to say that *s has* an acceptance system or that a set of propositions is the acceptance system *of* a person. Rather we should say that an acceptance system *is true of* a person, or is *about* a person. There seems to be a tension here between your terminology and definition.

The second point concerns what it means for a proposition (or more exactly for an acceptance) to *compete* with another one. According to your definition, *c* is a competitor of *p* if and only if the assumption that *c* is true makes it less reasonable for *s* to accept that *p* than the assumption that *c* is true. This, however, does not define the symmetrical relation of *competition*, but the asymmetrical relation of one proposition (acceptance) *undermining* another one. Take for example the proposition, which I accept, that I sometimes dream that I see a computer. That undermines the proposition, which I also accept, that I presently see a computer. If these propositions (acceptances) are competitors, the relation should hold the other way too. But that I sometimes dream of seeing computers is not undermined by my belief that I presently see a computer. Therefore I suggest the following improvement on your definition of a 'competitor': *c* competes with *p* on the acceptance system of *s* if and only if it is more reasonable for *s* to accept *p* on the assumption that *c* is false than to accept not-*p* on the assumption that *c* is true. This requires that *s decides* between either accepting *c* as true and rejecting *p* as false or accepting *p* as true and rejecting *c* as false. Such a decision is *not* required when I believe I see a computer and also believe that I sometimes see computers in my dreams. The latter belief may, but need not give me a reason to change my mind, whereas having competing beliefs is always a reason to change one of them. My third point concerns the principle of trustworthiness which you explain as telling us that if I accept that *p*, then *p* is more worthy of my acceptance than not-*p*. But why is it that *p* is more worthy of my acceptance than of my denial? The answer cannot be: Because *p* is true. This would turn the principle of trustworthiness into the unacceptable principle: If I accept that *p*, then *p*. Nor will the answer do: If I accept that *p*, then *p* coheres with my acceptance system. This only raises the further question what makes a whole system more worthy of being my acceptance system than another. Perhaps we could say that for *most* (*many, a lot of*) propositions it is the case that if I accept that *p*, then *p*? Is that what you mean by trustworthiness?

LEHRER: Let me start with the last one. I am trustworthing in what I accept means that my acceptance of anything I accept is a worthy guide to truth in the matter. I think there is a connection between statistical truth frequencies and worthiness as a guide to truth. I do not think they are the same thing. To suppose that they are is something like the naturalistic fallacy. To say that one thing is more worthy or reasonable for you to accept than another is connected with statistical truth frequencies, but it is a basically normative judgement. It is not one which is reducible to a statistical one. It is a comparative evaluative relation. The objection arises, not because of a definition of 'more worthy than' or 'more reasonable than', but because of the notion of correction of a belief about trustworthiness. The relationship between truth, rationality and trustworthiness is a normative one and hence not one that is easily reduced to a frequency notion.

Concerning the second objection about competition: You made a constructive suggestion. I certainly have to think about that. I cannot do it that quickly. You may undermine a claim, though you do not undermine the proposition. The *proposition* that you are Brandl is not undermined by the proposition that I sometimes mistake one person for another, but it might undermine my *claim* that you are Brandl. In a certain way it also competes with it. It competes with my claim, probably just in the sense, that it undermines it. But I am not quite clear why I should want to make that shift to your notion to obtain symmetry. The one I have got seems right, though asymmetrical, to me in the sense that, what I wanted to capture in the notion of a competitor is what constitutes an objection to my *claim*. So, I claim that I mean what constitutes an objection to it. There are other notions of undermining a competitor worth explicating, however. I do want to take a look on the definition of competition that you suggested. It strikes me as interesting.

FRANZ VON KUTSCHERA

REASONS AND CAUSES OF BELIEFS

I do not speak about Belief but only of beliefs of the common variety. My talk takes me across a very wide and controversial field. It cannot, therefore, be more than a brief statement on the most important points involved.[1] My theses will sound rather apodictical, since I leave out most of the arguments. I shall discuss the following questions: Do beliefs have reasons? Do they have causes? Are rational explanations of beliefs compatible with causal ones? I shall close with some remarks on the hypothesis of a causal determination of mental events.

1. DO BELIEFS HAVE REASONS?

For an answer we first have to agree on an understanding of beliefs and reasons.

1.1 What are beliefs?

Two preliminary remarks: We first have to distinguish between the state of affairs that somebody (at a certain time) believes that a certain proposition holds and this proposition believed by him. In what follows we are concerned with beliefs in the first sense, with beliefs as states of believing, not with what is believed. Secondly, "belief" is a common label for convictions, surmises, suppositions, expectations, etc. We say that a person X believes that a state of affairs, p, obtains only if p has a higher (subjective) probability for X than non-p, but different concepts of believing demand different differences between these two probabilities. For the sake of simplicity I shall always think of probability one for p, i.e. I shall only speak of convictions. Some people presuppose that probability one is almost never conferred to non-tautological propositions, and that certainly has its advantages in many contexts. But almost nothing I shall say depends on convictions having exactly probability one. Only at one point I shall use standard epistemic logic for beliefs which results from probability theory by restriction to probability one. Otherwise taking beliefs as convictions serves only to avoid cumbersome distinctions.

Now, for our topic the essential question is this: Is assuming an activity, are we free to assume one thing or another, or is it a process, something that just

P. Weingartner (ed.), Scientific and Religious Belief, 27–45.

happens to us? In other words: Do beliefs arise from choices or are they just passively conceived? Impressions are passive. By an impression I do not mean a sensation, but what I have when it appears to me that something is the case. I cannot determine, whether I shall have or go on having a certain impression. My impressions are something given for me. I can influence them only indirectly by closing my eyes, for instance, or by looking somewhere else. Judgments, on the other hand, are acts. Not only uttering a judgment is an act, but judgments are something we accept or reject or refrain from making, that we form, arrive at, suspend or revise. Beliefs seem to be located somewhere between impressions and judgments. On the one hand I also form or modify my beliefs, on the other hand many of them just obtrude themselves upon me; I cannot believe anything I want to. Often our beliefs follow our impressions: If it seems to me, as if my tobacco pouch is lying on the table, that's what I believe, normally. The Stoics said that the step from an impression, that something is the case, to the judgment, that it is indeed the case, is an act of assent which they termed synkatathesis (adsensio), a decision to accept the impression as correct. Now in judgments we express our beliefs. Therefore we cannot say that in our judgments we are free, but not in our beliefs. Beliefs, then, are neither impressions nor determined by them, but are also formed by an act of assent, of synkatathesis. We often accept that things are as they appear to us, but not always. The oar seems to bend where it dips into the water, but we still believe it to be straight. Knowledge, furthermore, is true and well-founded belief. Coming to know, doubtlessly, is an activity – we have to observe, argue or acquire informations. This activity does not concern truth, but only arriving at the belief. Of course the activity in forming beliefs often is not very pronounced, as for instance in the case of simple observations. But it becomes apparent when we have to weigh the evidences or have to sort out a lot of informations. The belief, for instance, that there are only four basic forces in physics certainly does not just obtrude itself upon us.

There are some objections against understanding beliefs as products of an activity. I shall try to respond to some of them briefly. We certainly cannot believe what we want to believe. I cannot, for instance, just convince myself that I am Chancelor Kohl. But I cannot run as quickly as I want to, either, or along the ceiling, and yet running is clearly an activity. That our capacity to do something is restricted is no argument that it is not an activity, that in proceeding with it we are not free. In our judgments we doubtlessly have a certain latitude, and therefore also in our beliefs. What we have to believe is generally what we have to believe if we want to observe criteria of rationality or hold on to fundamental convictions.

For a piece of behavior to be a free action of the agent it is necessary that he could have done otherwise, but it is not necessary that it be premeditated. Assuming an act of assent for beliefs does not then imply that coming to believe something always is preceeded by considerations. Sometimes it is, but often not.

It might be said, furthermore, that the assumption of acts of synkatathesis is superfluous, at least if we do not talk about convictions, but about probabilities. Just as convictions probability assignments have to be coherent, however, and coherence is not something which results from the impressions themselves; it has to be construed and there are mostly different ways to do that.

Lastly it is not a consequence of the Stoic thesis that we can always clearly distinguish between impressions and beliefs. Since impressions are not just sensations but appearances about what is the case, they are not independent of our beliefs, but theory-laden and interpreted in the light of preceeding assumptions. Some optical illusions, for instance, vanish as soon as we recognize them as such. The immediately given as the raw material of our observations, still free from hypothetical elements and interpretations, does not exist. The difference between impressions and assumptions is often only gradual, but nevertheless relevant, just as the judge's request that the witness restrict himself to facts and leave aside his interpretations of them.

In what follows, I shall thus proceed from the view that arriving at beliefs is an activity.

1.2 Reasons for beliefs

Reasons are what we normally refer to when we explain actions. A rational explanation of an action of Joe gives an answer to the question why he acted the way he did. It states aims, interests or, generally, preferences of Joe and beliefs or expectations he had, and shows that, in the light of these beliefs, the action was indicated for attaining these ends, or, put more abstractly, that it was rational in the sense of decision theory. The presupposition is that Joe, in the situation referred to, had alternatives, so that he could have done otherwise. Why he didn't get up from a chair he was chained to, is not a question that admits of an answer by rational explanation.

A rational explanation of a belief held by Joe is an answer to the question, why he holds it. Why does he believe that it will rain soon? Because he has seen dark clouds approaching. Why does he belief that it was Mr. Smith that killed Mrs. Brown? Because he saw him entering her flat with a knife, or because someone told him that the fingerprints on the weapon were Mr. Smith's. As reasons for beliefs we thus give observations, informations or other beliefs. In general we might say: Reasons for beliefs are other beliefs, that support them. Impressions are grounds for beliefs only if we take them to be reliable, i.e. only mediately.

If the adoption of beliefs is to be amenable to rational explanation, there have to be relevant interests and preferences, too. So what constitutes the utility of a belief for somebody? Karl Popper has discussed this question for theories and has pointed out that we are interested in true, informative and testable theories – testability is a prerequisite for ascertaining empirical truth –, as well as in simple, comprehensive and unified descriptions of the phenomena. We might simply say that we are interested in true beliefs, that belief is truth-oriented. Since truth is not something we can grasp directly, independently of our assumptions, the only way to reliable beliefs is to proportion them to the evidence. The bias for grounded convictions is common to us all, and therefore preferences for beliefs have to be mentioned only if they are not just rational ones. If I trust somebody, for instance, this may explain why I do not believe him guilty although there are good reasons to do so, reasons that others accept as conclusive. I may still believe in his innocence, not because I have assigned a higher a priori probability to it than they did, but because of personal commitments or emotional ties.

For beliefs to have reasons and be amenable to rational explanation they have to arise from decisions, then. Reasons, after all, are something the agent has for doing as he did, something that guides his decision. Inversely, if beliefs are adopted by acts of assent, they are in principle amenable to rational explanations; there is such an explanation if the agent had good reasons to do as he did.

Finally I want to emphasize, that a rational explanation of an act only shows it to be rational in the subjective sense, in the light of the preferences and expectations of the agent himself, but not in an objective sense. What he does may be subjectively rational but in fact not suited to his interests, or his aims may be devoid of any real value. In the same way a person's belief may be subjectively rational as seen against the background of his previous assumptions, but quite irrational according to scientific standards.

2. DO BELIEFS HAVE CAUSES?

2.1 What are causes?

The answer to the question whether beliefs have causes depends on the concept of causation employed. There are essentially three theories of causation that all take their departure from Hume: Regularity theory, probability theory of causation (P. Suppes) and counterfactual analyses (D. Lewis), and of the latter two there are objective and subjective versions – in the former case objective and subjective probabilities can be employed, and in the last one objective or epistemic interpretations of conditionals. These theories may confer different truth values upon singular causal statements. They all have their shortcomings. This may be taken as an indication that our talk of causes and effects in everyday life and in science does not follow one general criterion, that there is not just one concept of causation but several ones exhibiting only a certain family resemblance.

To get an answer to our question we have to commit ourselves to one concept, however, and I shall choose the one I have developed in my paper "Causation" (1993a). There I explicate the following notion of causation: An event E causes another event E', if the occurence of E' was not guaranteed until E occurred. Before E actually happened it was, then, still possible that E' would not occur, but from E's beginning it was necessary that E' would occur if E would. The necessity referred to is neither a logical nor a nomological necessity but a timedependent one. In a world it is necessary at a time t, that the event E' occurs, if E' occurs, no matter what turn the history of the world will take after t.

The frame of the analysis of this causal relation are branching worlds that form treelike universes. A typical case of causation is shown by the following section of such a tree:

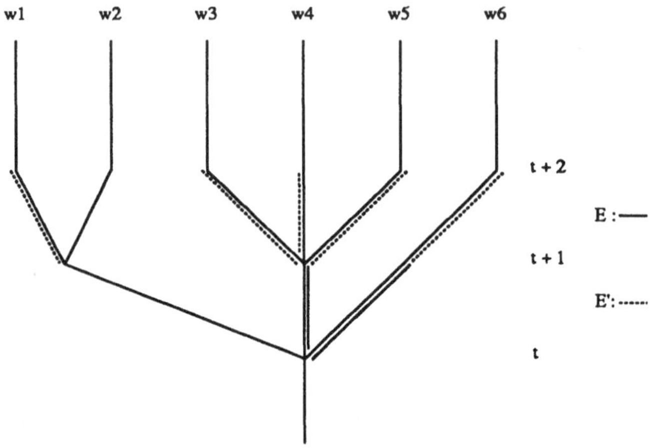

E occurs in the interval [t, t+1] in w3 to w6, E' occurs in these worlds as well as in w1 in the interval [t+1, t+2]. In w4 E causes E', since in t it is still possible that E' will not occur (see w2), but in t it is necessary that E' will occur if E does.

I think that this concept of causation corresponds to an important and frequently employed notion of cause, but it should be emphasized that there are some points in which it radically departs from other conceptions. In deterministic, unbranching worlds there are no causes, for instance, effects are never causes, since causes are only events that are not determined (i.e. necessary) from their beginning, while effects are always determined from the start. Hence there are no causal chains. This concept is related to that proposed by Georg Henrik v. Wright in "Causality and Determinism" (1974). Since I employ this notion of causation in what follows, it should be kept in mind, that my assertions depend on its specific properties. Causes, according to this notion, are typically chance events or actions, or events that contain them.

2.2 Do beliefs have causes?

A positive answer would have to come from the brain sciences, first of all, but as far as I see – I certainly don't see very far in these matters – they have little to offer in this line yet. So we have to look for everyday examples and argue on the phenomenological level. What we are interested in are, first of all, beliefs that cannot be rationally explained; the compatibility of rational and causal explanations will be our next topic.

Now I already said that a rational explanation of beliefs does not imply that they be rational in an objective sense. Unreasoned or unreasonable beliefs are non-starters in the present context, therefore. What remains are cases like these:

1) A child is afraid of dogs. Its fear will be bound up with the belief that dogs are snappish and impredictable. This conviction may have no reasons: The child has never been bitten by a dog, nor has it seen a dog bite someone else. Its parents assure it that dogs are quite harmless. There is a psychological explanation of this phenomenon as a projection of the childs fear of its father on animals. That sounds rather absurd, but still one would say in this case that the conviction cannot be explained rationally, but must have psychological causes.

2) A schizophrenic believes himself to be persecuted by dark powers. He can give no specific reasons for his conviction, and we should therefore say that it is an effect of his disease.

These are clearly border-line cases of beliefs. We are inclined not to call them beliefs but delusions. I don't want to deny that there are beliefs which can only be explained causally – the distinction between beliefs and delusions is terminologically not very well defined – but at least they seem to be atypical cases.

3. ARE RATIONAL EXPLANATIONS OF BELIEFS COMPATIBLE WITH CAUSAL ONES?

But cannot the same belief be explained rationally as well as causally? According to my notion of causation the answer must be in the negative: A rational explanation of a piece of behavior is only possible if the agent could have done otherwise. Only if he had alternatives we can attribute reasons to him for doing what he did. Whatever his interests or beliefs are, they would have been ineffective had he had no choice. If the action arose from a decision, however, it was not determined from its beginning and therefore it cannot be an effect of a precedent cause. So what can be explained rationally cannot also be explained causally. Inversely, effects are no possible explananda of rational explanations because they are not actions. What can be explained causally, then, cannot be explained rationally.

But cannot reasons themselves be causes? Can we not say: The fact that somebody has certain interests and convictions is the cause of his doing this or that? According to the concept of causation I am using here that is impossible even if we ignore the fact that interests and beliefs are not events. Even from the point of regularity theory we cannot accept reasons as causes. For causes are events at the beginning of which we can say: If they occur the effect will also occur. We cannot maintain, however, that if somebody has certain interests and expectations he will act in a certain way. We just don't always act rationally. If a sudden squall blows John's hat off we can explain his running after it by his desire not to lose it. But at the moment the hat is blown off his head we cannot say that John, since he does not want to lose it, will run after his hat he may well refrain from doing so since he does not want to make himself conspicuous. Causal explanations can be turned into predictions, rational ones not. Neither are causes reasons. A reason for adopting a belief is something that speaks for its truth. A cause for a belief, on the other hand, is neutral to its truth. In fact, if we recognize a belief as an effect of psychological or neurological causes that would not only be no reason to think it correct, but rather make us doubt it. Moreover, I myself can recognize a cause of a belief I presently hold only if I see it also as a reason for that belief: If I believe that an event E causes me to believe that A, I believe that E implies my belief that A. Now it is a theorem of epistemic logic that I believe my present beliefs to be true. Therefore I believe that E implies A, and hence my belief in E, which I must have if I take E to be a cause, is a reason for me to believe A.

Of course it is possible sometimes to show that the behavior of a person which can be causally explained, corresponds to his purposes, but that is no rational explanation of his behavior and does not show its purposiveness. If, for instance, somebody is thrown against a door in a train that suddenly slows down and it opens, just as he wanted it to do, that does not give us a rational explanation of an action.

4. MENTAL DETERMINISM

I have said: If the adoption of a belief is an act which on the phenomenological level at least is mostly the case it is amenable to rational, but not to causal explanation, and as far as beliefs have causes they are not results of decisions and therefore cannot be explained rationally. This still leaves open the question whether believing and thinking really are free activities or not. Do we, then, have to assume determinisms, the validity of the causal principle in the realm of the mental, or should we reject it?

As Kant has shown the question of universal determinism cannot be definitively decided. We cannot prove that every event has a cause, simply because we cannot produce an explanation for every event. If we produce such an explanation that is a new event for which we then have to find a cause, and so on. On the other hand, we cannot prove that a certain event has no cause, either, for we can never exclude that in the future we shall find a cause for it. We can only say that in view of the statistical character of the fundamental laws of physics universal determinism is probably wrong, or that a determinism of the mental should become probable, if the brain sciences should one day provide theories that allow for detailed explanations of mental phenomena of all kinds.

A complete determination of human behavior, especially of mental behavior is, then, at least logically possible, but I maintain that we cannot consistently believe that it holds. My arguments are not new and I cannot go into them here. One is the wellknown Gödel argument. It is agreed that it is not conclusive in the form it has been defended by J. R. Lucas. But it is correct in a version corresponding roughly to that of Paul Benacerraf in his paper "God, the devil, and Gödel" (1967), which does not attempt to show that man is not a machine, but that if he were a machine he could not recognize the fact. It may also be argued that we cannot possess a complete theory of human behavior. Such a theory would also have to describe how we understand and apply it, and that means that it would have to contain something like its own metatheory; the metatheory, however, is always richer than the theory itself. So there may be a theory describing and explaining human behavior down to its last details, but it is not a theory, which we can hold.

More important than such formal arguments, however, is to realize that our normal conceptions of, and with them our normal language about human behavior, are essentially bound up with the presupposition of freedom – freedom in the normal sense, not just in that of determinists like Hobbes and Spinoza, which Kant termed "die Freiheit des Bratenwenders", a "Bratenwender" being a grill operated by a clockwork. I we were to conceive of ourselves and others as automata, that would not only necessitate a far reaching revision of our conceptions of acting and thinking, but we would also have to develop a radically different language for expressing these notions. Presently at least, that is an utopian project in the sense of determinism even an absurd one, since according to it we just have to talk the way we do.

I want to illustrate this contention by two remarks: Communication by language presupposes freedom. According to Paul Grice a speech act is not only an intentional act but the speaker also performs it with the aim that his audience will recognize certain of his intentions. And the audience, in turn, understand the speech act only if they recognize the speaker's intentions. If speaking were a causally determined behavior, it would not be determined by intentions and could not, then, be recognized as intentional, still less as an attempt of the speaker to indicate his intentions. That, essentially, is an objection John Searle has raised in his book "Minds, Brains, and Science" (1984) against Alan Turing's thesis, that computers can talk. Searle insists that their outputs of symbols, though meaningful in our language, cannot be taken as a communicative acts, i.e. as expressions of intentions.

Knowledge also presupposes freedom. It consists in judgments which, as we have seen, are free acts. The norms of rationality, furthermore, that we try to satisfy in our argumentations, make sense only if we can conform to them, but do not do so automatically. Finally, the confidence we have in our understanding rests on the conviction that we are able to test our assumptions and revise them, that we do not just think the way we have to, but can take account of evidences and reasons that speak for or against the truth of our assumptions.[2]

All this, as I have said, can only be short hints, of course, but a detailed discussion would require a separate talk for each of the arguments I mentioned.

NOTES

1 For a more detailed discussion cf. Kutschera (1993), especially § 3.4.

2 For an argument against the compatibility of a very weak form of materialism with what we must be able to believe about the physical world in order to understand it cf. Kutschera (1994).

REFERENCES

Benacerraf, Paul: 1967, 'God, the devil, and Gödel', *The Monist* **51**, 9-32.

Kutschera, Franz v.: 1993, *Die falsche Objektivität*, de Gruyter, Berlin.

Kutschera, Franz v.: 1993a, 'Causation', to appear in the *Journal of Philosophical Logic*.

Kutschera, Franz v.: 1994, 'Global supervenience and doxastic logic', to appear in the *Journal of Philosophical Logic*.

Searle, John R.: 1984, *Minds, Brains, and Science*, Harvard University Press, Cambridge/Mass.

DISCUSSION

SCHURZ: We are starting the discussion and I am asking for questions.

LEHRER: I wanted to say something about the distinction between movements and action. If my hand moves up, then it seems to me, that might be determined. It doesn't follow from that fact that my action of raising my hand is caused, maybe it is not. But the peculiarity here is that actions imply movements or contain movements as a part or constituent. So suppose someone would maintain in the case of action like raising my arm there is a component – my arm going up – which was caused. That seems very plausible, because there are chemical reactions of my body and movements of muscles and nerves. They do seem to cause my arm going up. It would be very hard to believe that my arm going up, right now, has no cause, for the physiologist could probably tell us a good deal about the causation of my arm going up. So it looks like an arm goes up and looks like that's caused, when I raise my arm. Similarly one might say, my concentrating on a subject and thinking about it is something like an action. But the thought just suddenly occurring into me is less like an action and may be caused. So it looks like that my arm going up is caused. And yet you want to say, that my raising my arm is not caused.

KUTSCHERA: If I raise my arm and can stop any time, it cannot be caused. If, on the other hand there is an event after which my arm has to go up, then it would be caused. But this is very unplausible, I think. If we assume a causal determination of the process, it would have to be some sort of piecemeal causation in the sense of David Lewis. But that is not the main point. Let's take the raising of my arm to be a short process. Then the question is: Is it up to me whether my arm rises, or does it have a sufficient cause? In the latter case, according to my model, it would not be an action, and we would not speak of my raising my arm. I do not doubt that physiologists can tell us a long story about what goes on in my brain, nerves and muscles during my voluntary movements, but I doubt that this story amounts to a causal account.

LEHRER: So you think, that I initiate some event as a cause, which is an early element of a causal chain which results in my arm going up, roughly speaking.

KUTSCHERA: As Kant said, we can start causal chains. This does not mean that I cause my own actions. Actions are not caused, but are causes of other events. I cause something or bring it about by doing something. It is, of course, a question of fact, whether there are actions in the sense of my model, just as there is the question, whether there are genuine chance events. My model is just an instrument for an analysis of actions and chance events, if there are such things.

HICK: Do we have any way of telling whether there are actions in this sense?

KUTSCHERA: Yes. The presupposition that in much of what we do we are free agents is deeply embedded in our normal conceptions of human beings and human behaviour, and our normal language is tied to these conceptions. Therefore determinism, just as phenomenalism, cannot coherently be expressed in the normal language. It has to develop its own language, and that is not just a matter of introducing a new vocabulary. Using such a language would rather be a process quite different from the activity of normal linguistic communication, as I tried to indicate with reference to Paul Grice's analysis of linguistic meaning. So we have every reason to stick to the traditional paradigm as long as a coherent new one is not in sight.

HICK: Yes, but of course if we are have reasons to think we are not free, and change our language accordingly, then changing our language presupposes that we *are* free!

KUTSCHERA: You have just indicated that we are apt to run into a certain inconsistency, if we believe that we are not free. If we recognize, that there is no freedom, we should change our language, but we would still see this as an activity.

HICK: Yes. But if we are determined, then none of these problems of inconsistency arises, because we are determined to be saying exactly what we are saying and thinking what we are thinking.

KUTSCHERA: Determinism, as the proposition that every event is causally determined, is consistent, just as the proposition that I do not exist. An affirmation of either proposition by me, however, would be a pragmatical inconsistency. I can only affirm something, if I exist, and a statement, in the normal sense of the word, is an act that is not caused but guided by the speaker's intentions. We cannot come to believe that there are no actions, since adopting a belief is an activity. We cannot have command of a theory describing all human action deterministically.

HICK: You are suggesting that it is self-contradictory to believe, that we are caused to believe that we are free.

KUTSCHERA: Not only that, but there are arguments, that come more or less from logic. If there were a theory describing human behaviour in all its details, it would also describe, how we handle and understand it. Therefore, it would have to contain its own metatheory. But the metatheory always has to be richer than the theory itself.

HICK: But supposing that we are determinists, but we do not put forward a theory of how it is that everything happens, but simply the umbrella theory that whatever happens is determined.

KUTSCHERA: Putting forward a theory would still be an activity. The trouble with determinists is this: They do not see that what they maintain applies to themselves. While in their deterministic worlds there can be no talking and theorizing in the normal sense, they behave as if they were exempt from these restriction, like higher Beings contemplating the world from outside.

HICK: Yes, I agree with you that we are obliged, on pain of some kind of self-contradiction, to assume that we are free.

BRANDL: I am not sure, why you think, that on your concept of causation causal explanation is incompatible with explanation in terms of reasons or rationality. I think, I can best explain my worries with the diagram on the blackboard. If event E occurs in world 4, then according to this diagram, E causes event E' to occur, and it is a sufficient condition. But the next question arises of course, why the world developes to become world 3, 4, or 5, and I don't quite see, why we should not answer the second why-question now. The answer to the first question, why does E' occur, is, because it was caused by E. But now we have the second why-question, why does the world go on to of kind 3, 4, 5 and when answering the second why-question, why should we not bring in reasons in case that E' is an action. So if event E for example is an observation and E' is a reaction to this stimulus, then different reasons the observer has in making his observation may account for what the difference is in his further behaviour in the worlds 3, 4 and 5. So I could easily imagine to combine this causation tree with an explanation in terms of reasons, which explains the diversions between 3, 4 and 5. Why not?

KUTSCHERA: The event E does not determine whether world 3, 4 or 5 is realized. Another chance event or action will be responsible for that.

LEHRER: I have two questions. Suppose that people maximize an expected value and suppose that is a law of nature. Now that theory is not really deterministic, because you can have more than one action that has maximal utility. So it is not a deterministic model. Is that theory inconsistent with your view of action?

KUTSCHERA: Well, not with the formal model, but with my conception of preferences and beliefs not as causes, but as reasons.

LEHRER: Okay, here is the second question. This is a little more disputatious. One might defend, in fact – I try to defend – a notion of "could have done otherwise", that says, that you have access to possible worlds, that are nomologically inaccessible. In other words: given their conditions, it is determined, that you are in say world w4. But there is the notion of can or could have – that is the notion connected with action – that allows that you can have access to these worlds, which nomologically are not possible, since they are excluded by causal conditions. That is a fancy way of describing an old-fashioned view, that analyzed "can" in terms of "if-then", but I think, one can give a little bit more plausible account of traditional accounts by saying, there might be an accessibility relation such that, even though it is determined that you are in world w4, another world is accessible. Of course you explain that in traditional words by saying that if you had chosen to do otherwise you would have, and so on. We need to be convinced that this modality, this modality of can or could, which is tied up with causality, is the same modality of can or could that we require in the case of action. And that I guess is something I am less convinced of.

KUTSCHERA: You would say, somebody could have done otherwise even if that would not have been nomologically possible?

LEHRER: Right.

KUTSCHERA: Well, that is not the idea in my model. All these words would in fact be nomologically possible. So, as you said, I take a hard line. A piece of behaviour of an agent X is an action of X only if it would have been really possible, actually possible for X to realize a different alternative. Why do we have to use words, possible words, that are not nomologically possible, to stick to determinism? Why stick to determinism? It's wrong anyway.

LEHRER: Oh, that's true. I think, it is wrong. I simply – even if the determinism is wrong – am rather suspicious of the view that we are only free in those cases, in which the action is not determined. And roughly for the first argument I gave. I think, that there are times in which it may be deterministically settled that I perform a certain action. There may be very strong motives to do it and no motives to do the opposite. So, if I am very hungry and I sit down and a meal is put in front of me, I am going to eat this. Certainly I could resist, it's that I don't want to resist, and moreover I think that somewhere in the jungle of psychology is a causal explanation for that. The situation in which I don't eat my meal is excluded causally. But this situation seems to me in some sense available; I can imagine that I could have resisted. But I was really quite hungry and there was this food there and I had no incentive to hold up, I mean, there was no reason in the world not to eat the meal, I mean, there was this beautiful ham, and salad ... (everybody laughing). You begin to understand the causal chain.

KUTSCHERA: I should say, as long as you could have refrained from eating, your eating was an action in my sense. But if you did not have the possibility not to eat, then it was not an action. My concept of an action is of course rather exclusive, human behaviour may often turn out to be no activity in this sense. Maybe it is the case, that we often think, we act freely, but in fact we don't. My model does not give an answer to any material questions.

LEHRER: I am inclined to think, that the food put in front of me and my perception of the food and what's going on in my body does cause me to eat, but I think, I could have refrained. I think that there is a notion of accessibility possible worlds, in which, so to speak, we can do what is nomologically impossible. But this is important in the case of belief as well, because it looks like there are cases in which some causal process, for instance sensory stimulation, does cause belief. It looks like that in some cases. So I look across the table and I see a woman. Well, the visual experience that I have causes that belief. In that case it looks like the belief is caused, and it looks like it's rational, it's based on a very good evidence.

KUTSCHERA: But you could doubt, if you wanted?

LEHRER: Sure, I could have doubted, but of course the visual configuration does in fact cause the belief. It would be very hard for me to think that it doesn't. So as a result of that, I am inclined to be a compatibilist. There must be a sense, in which the belief could be caused. Caused by sensory evidence, the best evidence even if I could have refrained.

KUTSCHERA: I would not call this causation, as long as you could have refrained from believing it. It is causation, if it was practically automatical, that you accepted the belief. But if you could have done otherwise, then you had a choice and then there was no cause in this sense. For if there was a cause, you did not have a choice. Of course, we often say, that there is a cause for something if we don't mean a "causal cause", but just that there were grounds for it. So if we say in your case, that the visual presence caused the belief, that would be quite right in a normal meaning of causing, but not in the sense of nomological causation.

WEINGARTNER: I want to continue here. You said, reasons for belief are impressions for other beliefs. So, for instance, when scientists believe in some hypothesis and they change it after some time. There could be experiments for instance, the results of which can be understood as other beliefs or other impressions and therefore reasons for changing the belief, i.e. the hypothesis. On the other hand one could imagine such cases where your definition of cause would also be satisfied in some way. So for instance Einstein in his biography said that the special criticism of Mach and also these very simple "Gedankenexperimente" concerning lifts and trains were essential for him to come to a new concept of a time. There are other examples. So the question is whether we could say this idea could not have occured to him without Mach's criticism and without his "Gedankenexperimente". Another example would be the question, what was essential for discovering the DNA by Crick and Watson. And they themselves said, it was an essential point, that they stopped drawing pictures on a two-dimensional sheet of paper, but they began to work with wire models. Because then they could see the different angles of the chemical bonds in order to bring this spiral together. And they said this was essential, because Linus Pauling was very close to this idea also. So one could also ask, does this not satisfy your definition of cause, or in other words, can a belief in a hypothesis be refuted or be changed according to you only by reasons for a different belief or also causally? The discovery that the angles are so important is of course not a material cause, but a complex of an idea and a trial which is bound to a concrete material model.

KUTSCHERA: First, a necessary condition for an action is not a cause. Second, a belief is something different from an idea occurring to me. That may have a cause. Many ideas occur to me, that I entertain maybe for some time and then discard, but never assume to be true. So that is something else. There may be beliefs, which cannot be rationally explained, but which have causes, psychological causes or whatever. But they are borderline cases. Since the subject cannot help but have them and has no reasons for them, they would be something like compulsory delusions.

WEINGARTNER: It's exactly the question, whether there could have been other choices, what could have been another choice instead of the wire model? But the essential point here is, what one would call the reason or – perhaps in some weaker sense the cause –, the essential point was to have a kind of model which gives us the possibility to see the angles and the connections in a 3-dimensional shape.

KUTSCHERA: There were at least two decisions involved, one to try out a 3-dimensional model, and one, on the basis of its merits, to adopt the belief that it adequately represented the real structure of DNA-chains.

WEINGARTNER: Yes, but my question is, whether in the main point (3-dimensional flexible model to make the importance of the angels visible) there really is an alternative.

LEHRER: It seems to me that what you might say is, that having an idea or coming to have the idea might be caused. But that is not the same thing as belief. Once you have the idea, you have to evaluate and decide whether or not to accept it, and at that point the optionality comes in.

SCHURZ: If I may put a question as a chairman, my question is not so much about your notion of reason, because I tend to accept it, but about the notion of cause. I wonder, whether it fits to causality as viewed from modern natural sciences. I am not thinking so much about probability laws in quantum mechanics, but about chaos theory and so on. Your notion of causality has the feature that although interdeterministic as a whole, so concern the entire system of worlds, it is deterministic concerning the events, which cause an effect. So if the cause is obtained, the event is necessitated to obtain. I wonder, whether this is really true for most relations between events. If it takes serious natural sciences, then lot of events, most of them strictly can't be deterministically predicted. Moreover, it is somewhat strange for me, because an event is a small system and it is not really closed. So the chances, that you can have a deterministic relation between small systems, which are not really closed, are much less than you have a deterministic relation between the entire whole system. So if you assume that at the level of worlds, of total states of a system, you have no deterministic relation, then I wonder why one should then assume to have deterministic relations between the events themselves. So it seems to me it is plausible to assume just the notion of causality to be probabilistic, also between events and as soon as you make this step, then you have a compatibility between explanation by reasons and explanation by causes. What do you think?

KUTSCHERA: If you entertain a probabilistic theory of causation and take the probabilities as subjective probabilities, then there are connections, of course, between reasons and causes. A reason for a proposition P would then be a proposition, that makes P more probable, and that would also be a cause for P. But that sounds very strange for me, I must say. The other point was, that I consider indeterministic, open worlds. There are several possibilities how the world can go on in the future. The cause, E, determines the effect in the sense that if it occurs, the effect will follow, whatever else happens. If the effect depends on other things, then E is not its cause.

SCHURZ: Just to give you an example of moving the arm. I don't think, that there is any deterministic relation between two events, really, one ending at some time and the other starting at the time immediately later. And there is also a biological level. There may always be some disturbance factors intervening after the cause time period has ended and you just can't exclude possibility.

KUTSCHERA: Well, if that were the case, then there would be no causes in my sense.

SCHURZ: So that seems to me a rational basis for assuming that humans are free. Look at an event. In principle it is determined between the whole past, which can reach it. This past is by Einstein's theory very thick, I mean, because causes may have a velocity of light. But according to quantum mechanics, it is infinite, because you have action over the distance. So – according to quantum mechanics – if you want to have an event, which really determines some other event, it must be the state of the whole universe. I mean, that's of course very big, but if you assume only the relativity theory, it must be also very big, include more than the earth. So, do you see what I mean?

KUTSCHERA: Well, this is a problem, it is not said that at every branching point there are only two or three possibilities. Lots of things happening simultaneously. Different agents, be different things and different parts of the world, different chance events.

In my model the relation of causation has three arguments: It holds between two events in a world. The whole of the past history of our world may therefore be relevant for an event being the cause of something. Also the whole world developing in a certain way in some interval may be the cause of some later event. The narrowness of my concept of causality comes in, as you said, by the abundance of branchings that would have to be assumed in applications of the model. Lots of chance events and actions may happen simultaneously in different parts of the world, and only if in all these possible worlds the event E is followed by E', is E the cause of E'.

SCHURZ: If you think of time as really scaled, then you have an infinite branching.

KUTSCHERA: The worlds need not branch at every point in time. The origin of numerous branches is rather that there are lots of agents in a broad sense – Mother Nature is an agent, too; she is responsible for chance events.

SCHURZ: May I just mention one sentence. I also see a small paradox. If you assume you have the choice between x and y, so x and y are actions. If you have only these two continuations then to do x or y is caused according to your model.

KUTSCHERA: What is caused is that the agent has the two alternatives, but not one of them. A storm may be the cause for the captain having only two possibilities, throw the cargo over board or lose his ship.

LEHRER: Let us try one last thought experiment and see what you think of it. Let's suppose that Paul Weingartner is rather what you and I imagine him to be, that is a person, who has rational beliefs based on reasons and so on. Now imagine another possible world, in which there is somebody like Paul Weingartner, call him 'Twin-Paul', but this world has the following peculiarity, that 'Twin-Paul' is such, that every movement of his body and every thought that occurs to him, is caused. It's a deterministic world. But exactly the same movements and exactly the same thoughts are attributed to 'Twin-Paul' as we would attribute to Paul. And any moment, at which a certain thought goes through Paul's mind, 'Twin-Paul' in this other possible world has the same thought.

Two questions: Are you really ready to argue that 'Twin-Paul' performs no actions and has no beliefs? Second question: If you are willing to say that, then do you still think, that there is something rational about 'Twin-Paul' and something rational, say, about 'Twin-Paul's' thoughts?

KUTSCHERA: I should indeed say that 'Twin-Paul' performs no actions. He may have beliefs, for I said, that some beliefs may also be caused, but I should think, that we could not explain the behaviour of 'Twin-Paul' rationally, for what use would it be to refer to his preferences and expectations if they were not effective?

LEHRER: One thought might cause the next thought, that's what is bothering me. So if I have a thought, which is the premise of modus ponens and the next thought is the conclusion of modus ponens, there is something rational about that as opposed to someone who thinks about the same two premises and then arrives at the denial of the conclusion. One is rational, the other is irrational. Even if we suppose, that the one thought causes the next thought occurring, that the person could not refrain.

KUTSCHERA: I should say that the person whose "thoughts" are causally determined as the processes in a computer would behave as though he were rational, but he is not; his conclusions have no reasons, but causes.

GERHARD SCHURZ

NONMONOTONIC REASONING AND CHANGES OF BELIEF

1. FROM PHILOSOPHY TO ARTIFICIAL INTELLIGENCE

Deductive reasoning is truth-preserving reasoning, that is, a deductive inference from a premise set Prem to a conclusion Con, Prem $\vdash$ Con, is valid if and only if it holds in every possible situation, or 'model' respectively, that whenever the premises are true, the conclusion is true. As a consequence, deductive reasoning is *monotonic*, that is, whenever Prem $\vdash$ Con is valid, then also X, Prem $\vdash$ Con remains valid for every possible new information expressed by a sentence X, may X be true or false. So the validity of a deductive inference is independent from any possible information not contained in the premise set, and hence independent from the context of background knowledge. In this sense, the validity of a deductive inference is a 'local', or a 'modular' notion, and not a 'global', or a 'holistic' notion.[1]

In empirical science, one typically reasons from given scientific laws and facts to other facts. (There is of course the second important kind of reasoning from more general laws to more special laws; but we will focus here on this first kind of reasoning.) As a result, whenever deductive reasoning is applied as a model of scientific reasoning, as for example in Hempels deductive-nomologic (D-N) explanation model (Hempel/Oppenheim 1948), the laws used in the premise set must be *deterministic*, in the simplest case having the form $\forall x(Fx \rightarrow Gx)$, as in "every raven is black". For if the law $\forall x(Fx \rightarrow Gx)$ would not be understood as deterministic, but as admitting exceptional cases (that means, it could be true although admitting exceptional cases), the inference "$\forall x(Fx \rightarrow Gx)$, Fa/Ga" would be not truth-preserving and thus not valid. For a long time, inferences of the form "$\forall x(Fx \rightarrow Gx)$, Fa/Ga" were considered as a paradigm kind of scientific reasoning.

Deterministic laws and deductive reasoning play an important role in theoretical and idealized science. The reasoning patterns used here are typically mathematically more complex than inferences like "$\forall x(Fx \rightarrow Gx)$, Fa/Ga". In empirically applied science, however, as well as in ordinary reasoning of practical life, the laws are typically nondeterminstic, and as a consequence, the reasoning is not monotonic.

P. Weingartner (ed.), Scientific and Religious Belief, 47–76.

Consider some examples, like:

(1) This animal has wings, so it can fly
(2) This match is being struck, so it will light

The laws to which these two reasoning patterns implicitly refer are nondeterministic. There are some special, and rather exceptional cases of birds having wings which can't fly, like ostrichs, or penguins (their wings figure as flippers). Or, the wings of a bird might be broken. But in the *normal*, or usual case, animals having wings can fly. Similarly, a match might be wet, or even, it might be struck in an atmosphere which contains no oxygen. But again, in the normal case a match which just has been struck will light. So these laws can be stated as follows:

(1*) If an animal has wings, it normally can fly.
(2*) If a match is being struck, it normally will light.

Now, the basic idea of nonmonotonic is this: reasoning patterns like (1) and (2), implicitly refering to (1*) and (2*), are perfectly reasonable cases of reasoning, provided *nothing else* is known which constitutes a counterevidence to the conclusion and thus *blocks* the inference. Assume a biologist discovers animals which presumably are members of a so far unknown species, and these animals have wings. Then what she or he will conclude prima facie is that the animal can fly. Maybe the prediction later turns out to be false, due to special reasons discovered later. But at the present time, where the biologist knows just that this animal has wings, and nothing else which speaks against the conclusion, it is the best for him to draw the conclusion, that is, to *detach* the conclusion, that the animal can fly. Similarly, as long as we don't known that the match is wet, or that oxygen is missing, but only that the match is being struck, it is the best for us to conclude that it will light.

Let us formalize a qualitative nonderministic law like in (1*) and (2*) by $F(x) \Rightarrow G(x)$. We omit the quantifier, that is, all free variables are assumed to be universally quantified, with quantifiers in prenex form. Then the basic claim of nonmonotonic reasoning is that a reasoning pattern like

$Fx \Rightarrow Gx$, Fa therefore Ga

is correctly applied, that is, the conclusion *is detachable*, if only nothing else is known which may 'block' the inference. This reasoning pattern is clearly nonmonotonic. One cannot conclude from the correctness of the above inference that also "Fx $\Rightarrow$ Gx, Fa, Ha therefore Ga"is correct, for Ha may be one of the factors which block the inference - in our examples for instance the knowledge that this animal is a penguin, or that the match is wet. Similarly, the law "Fx $\Rightarrow$ Gx" itself is nonmonotonic in the sense that its truth is not preserved when strengthening its antecedent ("normally, a match will light if it is struck" is true; but "normally, a match which is wet will light if it is struck" is false"). Since the correctness of this kind of reasoning pattern implicitly refers to *what else* is known, to the background knowledge, it is implicitly holistic.

Several philosophers have discovered these kind of practical inferences already decades ago. For instance, in his arguments against the applicability of Hempels D-N explanation model to historical science, Scriven (1959) has mentioned that in most cases of empirical reasoning, one does not really use deterministic laws but laws which admit exceptional cases, which he called 'normic' laws. However, most philosophers at that time did not make much sense of the idea of normic statements. This is shown also by the fact that Scriven himself used to call them "truisms". The most frequent argument against these 'truisms' was that as long as we don't know what exactly is meant with the phrase "normally", these normic laws are immune against possible falsification, and thus contentless - that is, whenever the antecedent is true and the consequent is false, we just have to consider this as an anormal case.[2] Indeed, Scriven himself did not consider such 'truism' as proper scientific laws. This has led him to the claim that in history as well as in practical life, we have explanations *without* covering laws. But as I will argue below, normic laws, if properly understood, are a genuine kind of scientific laws, which may be called *qualitative inductive laws.* The clause "under normal conditions" will be interpreted in a precise sense - but the point will be, that it will be interpreted in an epistemic and not in an ontological sense.

The dominant attitude in the philosophy of science of the 60es and 70es, endorsed for instance by Hempel, was that there are only two strategies by which a 'normic' law can be transformed into a truly scientific law. Either the phrase "normally" is replaced by a complete list all further causally relevant conditions, which in their conjunction are sufficient for the consequence event, or the 'effect'. That is, the 'normic' law is completed such that it becomes a truly deterministic law. For instance "if a match is struck, and it is not wet, and there is enough oxygen, and the temperature is high enough (and...?), then it will light". However, this strategy is not possible in most practical cases, just because we don't know all these missing additional conditions; and even if we know them in principle, we don't know which of them have been present in the given case of application.[3]

The second strategy proposed by Hempel (1965) in his famous inductive-statistical (I-S) model of explanation was to replace the normic law by a probabilistic law. So we would have "the probability that a match will light, given it was struck, is so and so (say 90%)", formally $p(Gx/Fx) = 0{,}9$. The first important thing to mention here is that this probabilistic analysis gives us all the effects of non-monotonicity and context dependence as we obtain it in non-monotonic reasoning. Given a true statistical law $p(Gx/Fx)=0{,}9$ and the true antecedent Fa, we are only allowed to detach the conlusion that the inductive probability (that is, the rational degree of belief) of Ga is 0,9, if nothing else is known which 'blocks' the inference. For instance, if Fa stands again for this match being struck, and Ga for its lightning, and we know in addition that the match is wet, call this information Ha, then we are not allowed to conclude that Ga is true with inductive probability 0,9, because we know in addition the statistical law $p(Gx/Fx \wedge Hx)$ = low (say, 0,05). Hempel was fully aware of this phenomenon, which he has called "the ambiguity of I-S explanations", and he drew the conclusion that the correctness of such an explanation is dependent on what else is known in the background knowledge.

So indeed, in Hempel's I-S model the typical phenomena of nonmonotonic reasoning were already contained, with the only difference that he employed probability values, whereas nonmonotonic reasoning reasons without them. But this is the decisive point. For in most cases, we really don't know the probability values which are needed to reason probabilistically. For instance, we just don't know the probability that a match will light, given it is struck. It would be even ridiculous to undertake a statistical examination of this question, that is, to collect a random sample of all matches under varying conditions, and strike them all. What usually is done in applied statistical science is to say that there is a high statistical correlation between two events. But note that a correlation does not tell us any definite probability value. I claim that what is called "correlation" in applied statistical science (e.g., in empirical psychology or sociology) is usually evaluated not in a quantitative sense of probability theory, but in the sense of the qualitative inductive laws above. I also claim that nonmonotonic reasoning has no principial advantages over probability theory; its advantages lie in its low complexity and practical feasability, which allows efficient reasoning in situations where not only deterministic knowledge, but also statistical knowledge is missing.

Another field where philosophers (in particular, Rescher 1964) have discovered a similar phenomenon was the field of counterfactual conditionals.

They are also nonmonotonic (if this match had been struck, it would light" is true, but "if this match had been struck and had been wet, it would light" is false). According to the possible world analysis of counterfactuals, the claim that this match would have lit, if it had been struck, is a claim about a possible situation, or world, in which this match has been struck, and the claim implies that in this world, the match will light. But since the relation between striking a match and its lighting is not deterministic, there are several different possible worlds in which this match has been struck, in some of them the match will light, but in some of them it won't because, for example, in them the match is wet, and so on. So the question is in which of these possible worlds the truth of the counterfactual should be evaluated, and the answer, given by Rescher and others was, roughly, that "if A were the case, then B would be the case" is true, if B is true in that world in which A is true and which is otherwise as 'similar' as possible to the actual world. In this notion of similarity between worlds it is implicitly assumed that two world differing from each other only in an ordinary event, like striking this match, are more similar than worlds differing from each other in a normal condition. This assumption leads to the desired result that "if this match had been struck it would have lit" comes out as true, whereas "if this match had been struck, it would have been wet" comes out as false.

What is important for our purpose is that also the analysis of counterfactuals assumes a distinction between what normally is the case and what is not. So one could suggest that the above nonmonotonic reasoning patterns could be analyzed with help of counterfactual conditionals and similarities between possible worlds. Indeed this is a possible way. However, what we will suggest below is that in the opposite, counterfactuals can be analyzed with help of nonmonotonic laws and reasoning patterns. The advantage of this second way again does not lie in principal facts, but in matters of complexity and practical feasability. In practical situations, we are mostly unable to give a complete description of all the relevant possible worlds and their similarity relations. But we *are* able to state the basic qualitative inductive laws and the principles how to reason from them.

Though philosophy has procuded many insights which pointed towards the need of establishing a method of nonmonotonic reasoning with help of qualitative inductive laws, the decisive steps in this direction came from artificial intelligence. In what follows, I give a simplified reconstruction of the nonmonotonic logic as developed by McDermott and Doyle (1980), and in a somewhat different way, by Reiter (1980). The reconstruction is intended to point out one main problem in these first developments of nonmonotonic logic. I will then propose an improved model of nonmonotonic reasoning based on ideas of Nute (1988).

The knowledge system K, formally explicated as a set of sentences of a first order predicate logic including a modal operator, is divided into its *base* B $\subseteq$ K and into the set of *derived* consequences **Der** $\subseteq$ K. The base B contains those facts which are known by evidence, and the known laws, which may be either deterministic or non-deterministic. The set **Der** contains the facts which are derived from the base, either monotonically or non-monotonically. Since we reason from a given set of laws in the base, the laws figure like "rules" between facts. This is why in artificial intelligence laws are usually called "rules". In what follows, we shall adopt this terminology.

According to the basic idea of McDermott/Doyle and Reiter, the "normal" clause in nondeterministic rules is explicated in an epistemic way, meaning that "the consequent is possible according to our given knowledge". So, these rules can be explicated as[4]

$Fx \wedge \Diamond Gx \Rightarrow Gx$, for example
if x has wings and it is possible that it can fly, it can fly
where $\Diamond Gx$ is true *about* K iff $\neg Gx \notin K$ holds.

(Here and in what follows, the predicate variables F, G... may also stand for *complex* predicates, that is, as formulas containing x as a free variable).

Two things are remarkable about this reconstruction. First, the nondeterministic rule contains an antecedent clause which refers to the background knowledge K which contains the rule as an element. For this reason, Moore (1985) has called these rules *autoepistemic*. Second, because of this autoepistemic reconstruction, these rules seem to be monotonic. That is, given $Fx \wedge \Diamond Gx \Rightarrow Gx$ is in K, then whenever Fa and $\Diamond Ga$ are in K, it is allowed to detach Ga, whatever else is in K. For, if a counterevidence Ha is added to K which implies $\neg Ga$, then $\Diamond Ga$ is not longer acceptable in K, thus the antecedent of the rule is no longer instantiated. So, the rule of Modus Ponens applies to these reasoning patterns; the conclusion *is* detachable. But this monotonicity holds only at a superficial, syntactic level. Semantically, whenever a new information is added to K, the autoepistemic clause "$\Diamond$..." refers to a different background and thus changes its meaning - and it is *for this* reason that if counterevidence is added to K, then $\Diamond Ga$ becomes false and hence Modus Ponens is *not* violated. If $\Diamond Ga$ would maintain its old meaning, and thus would maintain to be true, we would have a counterinstance to Modus Ponens. So, autoepistemic rules are monotonic at the syntactic level (whence they allow an easy formalization), but non-monotonic at the deeper semantic level. Therefore, we propose to call them *quasimonotonic.*

Given a base B, let C(B) be the nonmonotonic closure of B. So K = C(B). So **Der** = K-C(B). Let us describe an algorithm of nonmonotonic reasoning, that is a procedure to reason from B to C(B). Given any set S, let the procedure **Pos** add to S *some* set of sentences of the form ◇Ga which are not yet in S, such that ¬Ga is not in S and Fx∧◇Gx ⇒ Gx is in S, for some F. Let the procedure **Con** add to S every sentence of the form Ga such that Fx∧◇Gx ⇒ Gx, Fa and ◇Ga are in S for some F. So the procedure **Pos** adds autoepistemic clauses which occur in rules of S and are true about S, and the procedure **Con** adds the consequences of the nondeterministic rules to S. It might be thought that the desired nonmonotonic closure C(B) is just the *fixpoint* of the iterative application of the procedures. This is only true in some examples, as in

B = {Fa, Fx∧◇Gx ⇒ Gx}
Pos(B) = {Fa, Fx∧◇Gx ⇒ Gx, ◇Ga}
Con(**Pos**(B)) = {Fa, Fx∧◇Gx ⇒ Gx, ◇Ga, Ga} = a fixpoint.

Consider the set B = {Fa, Ga, Fx∧◇Hx ⇒ Hx, Gx∧◇¬Hx ⇒ ¬Hx},
where for instance Fx = x is a match and it has been struck, Gx = x is wet, and Hx = x will light.
Assume **Pos**(B) = {Fa, Ga, Fx∧◇Hx ⇒ Gx, Gx∧◇¬Hx ⇒ ¬Hx, ◇Ha, ◇¬H}. Then
Con(**Pos**(B))={Fa, Ga, Fx∧◇Hx ⇒ Gx, Gx∧◇¬Hx ⇒ ¬Hx, ◇Ha, ◇¬H, Ha, ¬Ha},
which is *inconsistent.*

In this case, we have to perform a third operation, called retraction **Ret**, which – after each step **Con** – eliminates all autoepistemic clauses in S which are inconsistent with some facts in S, as well as those facts in S which have been derived with help of these autoepistemic clauses. So, our algorithm has to store the *dependency relations* between sentences in K (as in the "truth maintenance system" of Doyle (1979), a predecessor of nonmonotonic logic). So, our algorithm consists now in the iterative application of the procedures **Pos**, **Con** and **Ret**, until we arrive at a fixpoint. But the *problem* is that there exists not always a unique fixpoint; we may arrive at different fixpoints, or even at no fixpoint at all. In our example above, our algorithm **Pos-Con-Ret** may take three different ways:

Given B = $\{Fa, Ga, Fx \wedge \Diamond Hx \Rightarrow Hx, Gx \wedge \Diamond \neg Hx \Rightarrow \neg Hx\}$.

First way: add $\Diamond$Ha and proceed:
B_1:=**Pos**(B) = $B \cup \{\Diamond Ha\}$
B_2:=**Con**(B_1) = $B_1 \cup \Diamond \{Ha\}$
B_3=**Ret**(B_2) = B_2
Pos(B_3) = B_3=B_2. So B_2 is a fixpoint.

Second way: add $\Diamond\neg$Ha to B and proceed:
C_1:=**Pos**(B) = $B \cup \{\Diamond \neg Ha\}$
C_2:=**Con**(C_1) = $C_1 \cup \{\neg Ha\}$
C_3:=**Ret**(C_2) = C_2
Pos(C_3) = C_3=C_2. So C_2 is a fixpoint.

Third way: add $\Diamond$Ha and $\Diamond\neg$Ha and proceed:
D_1:=**Pos**(B) =$B \cup \{\Diamond Ha, \Diamond \neg Ha\}$
D_2:=**Con**(D_1) = $D_1 \cup \{Ha, \neg Ha\}$ = $B \cup \{\Diamond Ha, \Diamond \neg Ha, Ha, \neg Ha\}$
D_3:=**Ret**(D_2) =B
D_4:=**Pos**(D_3) = D_1
D_5:=**Con**(D_4) = D_2
D_6:= **Ret**(D_5) = D_3
... ad infinitum. So this way never arrives at a fixpoint.

The procedure given above is a simplification of that given in McDermott and Doyle (1980) or in Reiter (1980), but it leads basically to the same results.

2. FROM ARTIFICIAL INTELLIGENCE TO PHILOSOPHY

Summarized, the problem in this kind of nonmonotonic logic is that the notion of a "fixpoint", and thus, the notion of "nonmonotonic consequence", is ambiguous. Both MacDermott and Doyle and Reiter suggest solutions to overcome this difficulty. The former suggest to take the intersection of all fixpoints as the set of 'unambiguous' nonmonotonic consequences. The latter suggests to take the disjunction of consequences in different fixpoints. I think that both suggestions are not satisfying. In the example above, both suggestions would lead to the result that there is no new nonmonotonic consequence (because the intersection of fixpoints contains no new consequence, and the disjunction is empty). However, if we know that this match has been struck and it is wet, we surely will conclude that it will not light, that is, we will prefer the second rule "normally, a match which has been struck and is wet will not light" over the first rule "normally, a match which has been struck will light", because the second rule has a *more specific* antecedent that the first rule. So what we need is a preference rule between rules, which, whenever we are in a situation where we have two rules with instantiated antecedents but conflicting consequents, tells us whether any of them fires, and if so, which one.

Exactly such a preference rule has been explication, in various versions, by Hempel and his followers in the inductive statistical explanation model. It was called the condition of *maximal specifity*, according to which an I-S-pattern is correct, or – in our terminology, a nondeterministic rule fires – only if its instantiated antecedent contains the most specific information about the actual situation, among all instanciated antecedents of rules. Nute (1988) first made an attempt to construct an unambiguous nonmonotonic reasoning system embracing a condition of maximal specifity. In what follows, we give another reconstruction of this idea, which differs from Nute's reconstruction mainly in the fact that our reconstruction will be autoepistemic in the mentioned sense, that is, the maximal specifity condition will be part of the antecedent of the nondeterministic rules, as in the nonmonotonic logic of McDermott and Doyle.

We explicate nondeterministic rules in the form

$$Fx \wedge \text{Not-Blocked}(\langle Fx,Gx \rangle,x,K) \Rightarrow Gx.$$

We call Fx the *ontologic antecedent part* and Not-Blocked(<Fx,Gx>,x,K) the *autoepistemic antecedent part.* Note that the "x" occurring in "<Fx,Gx>" has to be understood as bound (alternatively, we could also use quotation marks, that is, we could write "<"Fx","Gx">), whereas the x occurring as the second argument of "Not-Blocked" is free and is instatiated by the same individual name as that which instantiates the ontologic antecedent part. Not-Blocked(<Fx,Gx>,a,K) means that given Fa is in K, the application of this rule to the individual a is *not blocked* by the knowledge in K, and so, the rule *fires.* To simplify the terminology, we write our rules in the form

$$Fx \wedge (!) \Rightarrow Gx$$

where (!) just abbreviates "Not-Blocked(<Fx,Gx>,x,K)". So the notation "(!)" is a *twofold contextual* one: its meaning depends on the rule in which it occurs, and on the knowledge system in which this rule occurs.

We will also admit deterministic rules in K, which we abbreviate by "$Fx \rightarrow Gx$" (to be read as universally quantified). For reasons of simplicity, we assume that K is closed under deterministic rules, that is, every deterministic rule implied by K's deterministic rules is itself in K.

If a is any individual, we define Not-Blocked(<Fx,Gx>,a,K> to be true iff

(*) there *is no sentence* of the form Ha in K such that
(a) either Hx is the formula $\neg Gx$ itself, or $Gx \rightarrow \neg Hx$ is in K; or
(b) either a deterministic law of the form $Hx \rightarrow Qx$ or a nondeterministic law of the form $Hx \wedge (!) \Rightarrow Qx$ is in K, where $Gx \rightarrow \neg Qx$ is in K, and $Fx \rightarrow Hx$ is not in K.

Clause (a) excludes the case when the negation of Ga or a sentences deterministically implying the negation of Ga is already in K, for example because of evidential knowledge. Clause (b) excludes the case where we have another instantiated rule, deterministic or nondeterministic, which is *not less* specific, and which has a conflicting consequent.

Some clarifications about the logical nature of this kind of reasoning pattern are appropriate. First, it is quasimonotonic in the above sense and we can apply the rule of Modus Ponens. This does not mean, of course, that this kind of detachment is truth preserving. It may well happen that Fa and $Fx \wedge (!) \Rightarrow Gx$ is true, but $\neg Ga$ is false; simply because the rule is non-deterministic. A rational requirement for such a rule to be true, or rationally acceptable, is certainly that the conditional probability p(Gx/Fx) must be high, at least greater than a half (and moreover, greater than the prior probability of Fx). So it can never happen that both $Fx \wedge (!) \Rightarrow Gx$ and $Fx \wedge (!) \Rightarrow \neg Gx$ are rationally acceptable. However, since nonmonotonic reasoning reasons without probability values, it can't be said how great the value of p(Gx/Fx) is, and also, to what extent these probability values are preserved or not preserved in chains of nonmonotonic reasoning steps. An investigation of this question would certainly be important, but it lies out of the scope of this paper.

Detachability in nonmonotonic reasoning does not preserve truth, and we don't know up to which extent it preserves probability, but at least, it preserves consistency: if we apply one step of Modus Ponens to a consistent set of facts, autoepistemic clauses and rules, the derived facts are also consistent. Since after each step of Modus Ponens we obtain new derived facts, we must update the autoepistemic clauses, before we apply Modus Ponens again. Otherwise, inconsistent results would be possible (for an example see below). So the algorithm to obtain a unique fixpoint of derived consequences from a given base set B is this: we first apply the operation **Auto** which adds to a given set S all instantiated autoepistemic antecedent parts (of rules in S) which are true about S – and for the sake of iteration – which eliminates all instantiated autoepistemic antecedent parts which are contained in S but false about S. Then we apply the operation **MP** which adds to a given set S all nonmonotonic consequences derivable from some subset of S by one step of Modus Ponens. One can show that the result, **MP(Auto(B))** = B_1 is again consistent. We iterate the procedure **Auto** - **MP** until we arrive at a unique fixpoint, K = **MP(Auto(K))**. Of course, what I have given here was only a rough scetch; a detailed proof of the existence of a unique fixpoint and an investigation of some interesting variations of this algorithm has to be left to further studies.

Let us now consider how our reasoning patterns work at hand of some examples. Assume we have the following rules in B

(R1) Match(x)∧Struck(x)∧(!) ⇒ Lights(x)
(R2) Match(x)∧Struck(x)∧Wet(x)∧(!) ⇒ ¬Lights(x)

and we know (by evidence) that this match (a) has been struck, that is, Match(a), Struck(a) ∈ B. If we know nothing else of relevance, then rule (R1) fires, giving us Lights(a). Of course, should we have already observed that a, although it had been struck, did not light, then ¬Lights(a) is in B and (R1) is blocked, because of the clause (a). Should we in addition know that the match is wet, Wet(a) ∈ B, then again rule (R1) is blocked, but now, rule (R2) fires.

Now let us look at a situation of competing rules, where no one is prefered, as in

(R3) Has-Wings(x)∧(!) ⇒ Can-Fly(x)
(R4) Mammal(x)∧(!) ⇒ Can't-Fly(x)

where we assume, a is a bat, and we know about a that it has wings *and* it is a mammal, that is, Has-Wings(a) ∈ B and Mammal(a) ∈ B. Then *both* rules are blocked, which is reasonable, since we can draw no safe conclusion in such a situation. However, given we know that all bats normally can fly and that all bats are mammals, so given the rules

(R5) Bat(x)∧(!) ⇒ Can-Fly(x)
(R6) Bat(x) → Mammal(x)

are in B, and Bat(a) is in B, then the rule (R5) will fire, because "Bat" is more specific than "Mammal". Observe that (R6) is a deterministic rule, whereas (R5) is nondeterministic; for example, if a bat has broken wings, it won't be able to fly.

Let us finally discuss a special example in the context of interative applications of the procedure **MP**. Assume our base set is

$$B=\{Fx\wedge(!)_1 \Rightarrow Gx,\ Gx\wedge(!)_2 \Rightarrow Hx,\ Qx\wedge(!)_3 \Rightarrow \neg Hx,\ Fa,\ Qa\}.$$

We have indexed the different fire-conditions in the three rules. Then the procedure **Auto** first adds (a-instances of) $(!)_1$, $(!)_2$ and $(!)_3$. The procedure **MP** adds then Ga and ¬Ha. If we would now apply **MP** again, we would obtain an inconsistency. Instead, we have to apply **Auto** first, which now eliminates $(!)_3$. Now **MP** adds nothing more, and we have obtained the fixpoint. We have here a situation where the second rule 'indirectly' competes with the third rule, that is, it competes after having performed several numbers of **MP**, but since in the iteration of **Auto** and **MP**, the antecedent of the third rule is instantiated before the antecedent of the second rule gets instantiated, the third rule fires and the second is blocked forever. Here, a variation of our algorithm would be conceivable with the effect that in such a situation of an 'indirect' competition, both rules are blocked. However, the explication of this idea would certainly be no easy task, and maybe the uniqueness of fixpoints would then get lost in such an enterprise.

Let us show how some central notions in Keith Lehrers theory of belief systems, which to a great extent has been developed in the context of subjective probability theory (cf. Lehrer 1974), can be explicated within our framework.

Lehrer defines that A *competes* with B if it is more reasonable to accept B on ¬A than to accept B on A. Assuming A and B are open formulas, A(x) and B(x), this situation is given if we have accepted two rules of the form

(R7) $X(x)\wedge(!) \Rightarrow B(x)$
(R8) $X(x)\wedge A(x)\wedge(!) \Rightarrow \neg B(x)$

where X(x) is any (possibly empty) additional antecedent clause. If we admit contraposition as a valid rule for nondeterminstic laws, which seems plausible, then (R7) implies

(R9) $X(x)\wedge B(x)\wedge(!) \Rightarrow \neg A(x)$

and the notion of competition becomes symmetric.

Next, Lehrer defines that B *beats* A if A competes with B and B is more reasonable than A. This is the case if we have accepted the two rules above and in addition, X(a) is in the base set and A(a) is not in the base set. Then B(a) will be in our knowledge set (either in its base or derived), but not A(a).

Finally, he defines that N neutralizes A as competitor for B if A competes with B, but A∧N does not compete with B. This is the case if the two rules (R7-8) above are accepted and in addition the rule

(R10) X(x)∧A(x)∧N(x)∧(!) ⇒ B(x)

is accepted in K.

Next we want to point out that our reasoning pattern can be seen as a *logic of neurons.* Every such rule corresponds to a neuron, with several inputs and one output. Ontologic antecedent parts correspond to the activating inputs of a neuron, while the consequents of the competing rules to which the autoepistemic clause refers figure like inhibiting inputs of a neuron. For instance, the situation of the two rules (R7) and (R8) corresponds to the following neuronal pattern:

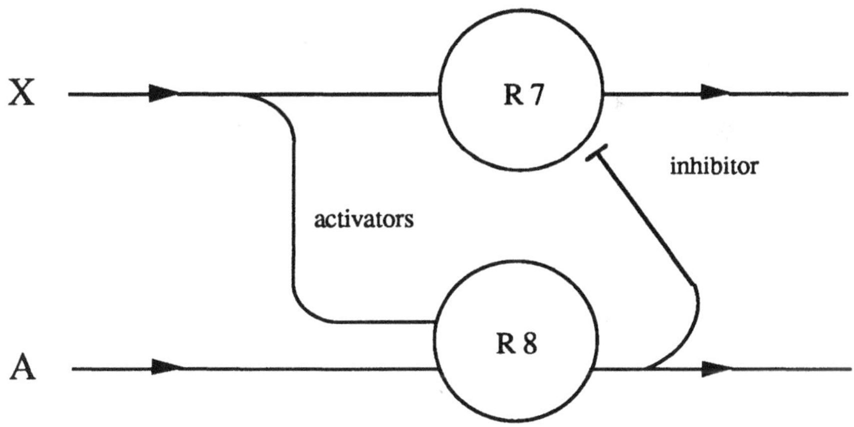

3. CHANGES OF BELIEF

Assume K is our belief system and A is a new information inconsistent with K. Then A changes K into a new belief system K*, which is called the revision of K by A. According to a widely accepted approach (cf. Gärdenfors 1988), K* is the result of first contracting K in a minimal way such that it gets consistent with A, and then expanding this contracted system, denoted as K-A, by A, so $K^* = K\text{-}A \cup \{A\}$. K-A is a maximal subset of K which is consistent with A. But the problem is that usually, there are several subsets of K with maximal cardinality, each consistent with A. Let us illustrate this point at hand of the example of striking a match. Assume K contains the following sentences

In the base:

M(a) this is a match

and the rule (which we cautiously formulate as nondeterministic)

(R11) $M(x) \wedge ST(x) \wedge \neg W(x) \wedge O(x) \wedge T(x) \wedge (!) \Rightarrow L(x)$.

In the set of derived assumptions:

$\neg ST(a)$ this match is not struck
$\neg W(a)$ this match is not wet
O(a) there is enough oxygen in the surrounding of a
T(a) the surrounding temperature is high enough (i.e., it is not very low)
$\neg L(a)$ this match does not light

Assume we receive the new information A:= ST(a) by evidence. There are now several possibilities to contract K, that is to form K-A. Of course we have to eliminate in K $\neg ST(a)$, but in addition, we must eliminate either one of the singular assumptions above, or we must eliminate the rule. Note that the situation is exactly the same, whether we are in the indicative mood, that is we consider an actual belief change, or we are in the counterfactual mood, and consider a hypothetical belief change, a 'swift' to another 'possible world' where ST(a) is true.

Of course in this situation we will eliminate $\neg L(a)$, and then infer L(a), so we will reason "this match has been struck, and so it will light", rather then, for example "this match has been struck, and so it was wet", or "this match has been struck, and so there was not enough oxygen"; and similarly, in the counterfactual mood. In the theories of belief change as well as in the theories of counterfactuals, this is usually modeled by assuming a preference relation between sentences (or alternatively, between sets of them), which in our case is

$$ST < \neg W < O < T < \text{a rule} < \text{an evidence}$$

ST is an ordinary event, while NW, O and T are 'normal conditions', with increasing 'normality' (or probability, respectively). So this preference means that we prefer to give up an ordinary fact assumption before we give up a normal condition, and we prefer to give up a normal condition before we give up a law. In our example of a belief change, this leads to the result that given the new evidential information ST(a) is added to the base B, whence ¬ST(a)is removed from **Der**, and nothing else is known in B, then ¬L(a) is removed from **Der** and we conclude L(a). If we, on the other hand, already know that the match did not light, i.e. ¬L(a) ∈ B, then we conclude that the match was wet, i.e. we remove ¬W(a), and add W(a); and so on. Note that in our refined model, where we distinguish in K between the base B and the set **Der** of derived facts, new informations are always assumed to be new evidence and thus are added to B, while contractions are usually performed in the set **Der**. Only in the case where a rule in B is falsified by a new evidence, or where an old evidence is falsified by a new one, contractions are performed in B.

The dominant position of an evidential fact is adequate only for the indicative mood, while in the counterfactual mood, actual evidence counts like derived facts. In other words, counterfactuals are evaluated independent from our actual evidence - we always have to fix what we definitively know in the hypothetical knowledge by stipulation.

As Peter Gärdenfors (1990) has recently shown, a belief set K figures like a nonmonotonic reasoning machine (a set of defaults in Reiter's sense), which, if certain new informations A are added to K, gives certain new consequences B, which behave nonmonotonically. He thus establishes a global connection between nonmonotonic logic and the theory of belief change. He gives, however, no hint how the structure and change of K can be modeled with help of nondeterministic rules. In what follows, we show how this can be done. More specifically, we will demonstrate how the above ordering relation can be replaced by a certain set of nondeterministic rules which altogether lead to the same results. In our example, these rules are:

(R12) $M(x) \wedge ST(x) \wedge (!) \Rightarrow L(x)$
(R13) $M(x) \wedge ST(x) \wedge \neg L(x) \wedge (!) \Rightarrow \neg W(x)$
(R14) $M(x) \wedge ST(x) \wedge \neg L(x) \wedge \neg W(x) \wedge (!) \Rightarrow \neg O(x)$
(R15) $M(x) \wedge ST(x) \wedge \neg L(x) \wedge \neg W(x) \wedge \neg O(x) \wedge (!) \Rightarrow \neg T(x)$.

The fact that we usually assume a match to be not wet is represented by the following rule

(R16) $M(x) \wedge (!) \Rightarrow \neg W(x)$

which states a "normal conditions for matches". That O and T are general normal conditions, that is, if we don't know otherwise, we usually assume them as true, can be modeled by the following rules the with empty antecedent clauses:

(R17) $(!) \Rightarrow O(x)$ (R18) $(!) \Rightarrow T(x)$

Let us assume that M(a) and (R12)-(R18) are in B, ¬ST(a) is in **Der** (while for the moment, we don't assume ¬La to be in **Der**). The new information added to B is ST(a). Then ¬ST(a) is contracted from **Der**. What else happens depends on what else is known in B:

(1) Assume nothing else of relevance is in B. Then rules (R16-18) fire and ¬W(a), O(a) and T(a) are in **Der**. So rule (R12) fires, and L(a) is added to **Der**.
(2) Assume in addition, ¬L(a) is in B. Then rules (R17-18) fire and O(a) and T(a) are in **Der**, but rule (R16) as well as rule (R12) are blocked by the more specific rule (R13), which fires and adds W(a) to **Der**.
(3) If ¬L(a) and ¬W(a) are in B, then rules (R12), (R13) and (R17) are blocked by the more specific rule (R14), which fires and adds ¬O(a) to **Der**, moreover rule (R18) fires and adds T(a) to **Der**.
(4) If finally ¬L(a), ¬W(a) and O(a) are in B, then rules (R12)-(R14) and (R17-18) are blocked, and only rule (R15) fires, adding ¬T(a) to **Der**.

We may assume in the whole story that we have the additional rule

(R19) $M(x) \wedge \neg ST(x) \wedge (!) \Rightarrow \neg L(x)$

that is, normally a match which is not struck does not light. Then the knowledge situation before adding ST(a) to the base is like before, except that also ¬L(a) is in **Der**. After adding ST(a) to B, we have then to eliminate not only ¬ST(a) but also ¬L(a) from **Der**, because the derived fact ¬**Der**(a) was dependent on the assumption ¬ST(a). We see that we have to store the dependency relations between sentences in B and in **Der**, in order to perform contractions of the set **Der** induced by new informations added to the base B. The rest remains then as discussed above.

It is clear that in this way, the nonmonotonic reasoning pattern gives us not only a theory of belief change, but also an alternative analysis of counterfactuals.

We finally want to point out an important difference between monotonic and nonmonotonic belief change. Under an expansion or contraction of a set, we always understand an operation forming a (proper or improper) superset, or subset respectively, of this set. In the monotonic case (where we only have deterministic rules), an expanasion of the base B always leads to an expansion of the derived consequences **Der**, and a contraction of B leads to a contraction of **Der**. In the nonmonotonic case, however, it may happen that an expansion of B leads to a contraction of **Der**, and a contraction of B leads to an expansion of **Der**. To give an example, assume

$B = \{Fx \wedge Qx \wedge (!) \Rightarrow Gx,\ Fx \wedge Hx \wedge (!) \Rightarrow \neg Gx,\ Fa,\ Qa\}$
so **Der** = {Ga}.

If we add Ha to B, we obtain

$B^* = B \cup \{Ha\}$.

We have to eliminate Ga from **Der**, and since no one of the competing rules is prefered, no one of them fires, so

Der* = {}.

To obtain an inverse example, we just have to put B* instead of B, and **Der*** instead of **Der**. Elimination of Ha from B* leads back to B, and we obtain **Der** as the new set of derived consequences, because the first rule is not blocked any longer.

NOTES

1 Why the above rule Prem ⊢ Con ⇒ X∪Prem ⊢ Con (for any X) is called the rule of monotonicity can be best understood from the theory of consequence operations, where one deals with the deductive closure Cl(Prem) of sentence sets instead of the deducibility notion. Here this law can be equivalently stated as
$\text{Prem} \subseteq \text{Prem}^* \Rightarrow \text{Cl(Prem)} \subseteq \text{Cl(Prem}^*)$
which means, that the closure operation behaves monotonic under the relation of inclusion between premise sets. A general set theoretical explanation of the term "monotonic" is this: given

a set A, an operation on A op: A→A, and a partial ordering relation R on A×A, then op is called monotonic with respect to R iff aRb implies op(a)Rop(b). In our case, R is ⊆ and op is Cl. The terminology comes from mathematical functions, where A is a set of numbers, R is the relation ≤ (which here is a total ordering), and op is a number function, which is called 'monotonically increasing'.

2 For an argument of this kind cf. Albert (1957).

3 A well known argument against the D-N model of explanation based on this consideration has been put foreward by Cahnfield and Lehrer (1963).

4 Note that this reconstruction of nondeterministic rules is in accordance with the one given by Reiter, while McDermott and Doyle allow any rules of the form $Fx \wedge \Diamond Gx \Rightarrow Hx$; moreover they write them as statements.

REFERENCES

Albert, H.: 1957, 'Theorie und Prognose in den Sozialwissenschaften', reprinted in: Topitsch, 1971, 126-143.

Cahnfield, J. and K. Lehrer: 1963, 'A Note on Prediction and Deduction', *Philosophy of Science* **28**, 204-211.

Doyle, J.: 1979, 'A Truth Maintenance System', *Artificial Intelligence* **12**, 231-272.

Gärdenfors, P.: 1988, *Knowledge in Flux*, MIT, Cambridge/Mass.

Gärdenfors, P.: 1990, 'Report of the ECAI '90', *AI Communications* **3**, 3, 146-152.

Hempel, C.G. and P. Oppenheim : 1948, 'Studies in the Logic of Explanation', reprinted in: Hempel (1965), 245-290.

Hempel, C. G.: 1965, *Aspects of Scientific Explanation and Other Essays*, Free Press, New York.

Lehrer, K.: 1974, *Knowledge*, Clarendon Press, Oxford.

McDermott, D. and J. Doyle: 1980, 'Non-Monotonic Logic I', *Artificial Intelligence* **13**, 41-72.

Moore, R.C.: 1985, 'Semantical Considerations on Nonmonotonic Logic', *Artificial Intelligence* **25**, 75-94.

Reiter, R.: 1980, 'A Logic for Default Reasoning', *Artificial Intelligence* **13**, 81-132.

Rescher, N.: 1964, *Hypothetical Reasoning*, North-Holland, Amsterdam.

Scriven, M.: 1959, 'Truisms as Grounds for Historical Explanations', Gardiner, P. (ed.), *Theories of History*, Free Press, New York, 443-468.

Topitsch, E. (ed.): 1971, *Logik der Sozialwissenschaften*, Kiepenheuer & Witsch, Köln, 126-143.

DISCUSSION

KUTSCHERA: We start the discussion of Dr. Schurz's paper. May I ask for questions?

LEHRER: I am puzzled by the following. It seems to me, that all is going to be blocked. For that reason as far as I can tell you are looking at the classes here extensionally and if you are, then you can always manufacture a predicate which will block the inference. You can always manufacture some class that's going to block the inference.

SCHURZ: How?

LEHRER: Well, assume you are trying to get an inference to A. Basically the problem is this. Suppose that there is some class C that leads to not-B. So A leads to B and C leads to not-B. Assume x is a member of A, x is not a member of C. This is a case in which the inference should go through. So what I am going to show you is, that it is going to be blocked anyway, because you can find a new class, which is the union of C and the set containing only x, $C \cup \{x\}$.

SCHURZ: We have to translate into formulas. The formula is $Cy \vee y = x$. I think it was an objection of Grandy against Hempel.

LEHRER: Yes, it is a standard objection. Call the predicate "$Cy \vee y = x$", C^*. So the extension of C^* is the class $C \cup \{x\}$. If C is a big class, one thing more or less will not make any difference. So the inference from A to B will not go through because of C^*. We get that the rule $C^* \wedge (!) \Rightarrow B$ holds.

SCHURZ: Why not?

LEHRER: If you have got an inference from A to B, membership from A to membership from B, then there is always going to be a blocker, if there is any set that leads to not-B, because you can artificially contrive a new set C^*, of which x is a member, that will lead to not-B.

SCHURZ: My reply is that, first, I think the debate about these acceptance conditions in probabilistic reasoning has shown that you need a criterion for a nomological predicate. And C^* is not a nomological predicate. I think, all the proposed criteria of maximal specifity involve the notion of a nomological predicate.

LEHRER: Oh well.

SCHURZ: With a non-nomological predicate defined like C^*, you can always undermine such a criterion of maximal specificity. I have to assume of course that these rules involve only predicates which are really nomological, defined in some reasonable intentional way, and not just any irrelevant mixtures like C*. I think this is an answer. One should not accept C* $\wedge$ (!) $\Rightarrow$ B as a rule, because it is not nomological.

LEHRER: But it seems to me, that you are going to want to allow blocking predicates, that are not nomological predicates, because sometimes I can ask for specific information about something, that does not involve any general law, but just involves some very specific piece of information. So for instance if something is a duck, then it can fly, but suppose, I have this specific information, that the ducks in Green Lake have their wings tied, so they cannot fly. Now being a duck in Green Lake is not a nomological predicate.

SCHURZ: Being a duck in what?

LEHRER: In Green Lake. That's a lake, what's called Green Lake, you may call it Gerhard's Lake. Anyway, assume you have put some ducks in there, but you don't want them to fly away. So you tied their wings. Now being a duck in Gerhard's Lake is not a nomological predicate.

SCHURZ: It's not very pleasurable to be in this lake.

LEHRER and others (laughing): Well, we don't want to assume, it's not good. But the problem is, it seems to me, there you want to use being a duck in Gerhard's Lake as to block the inference.

SCHURZ: Sure, in this case it is reasonable. If I understood you right, the rule is: if you are a duck and are in Gerhard's Lake, then you can't fly. But your previous example was a disjunction: "being C or being identical with x". Now you have a conjunction "being a duck and being in Gerhard's Lake".

LEHRER: Yes, but that is not a law-like predicate.

SCHURZ: Sure, but in this case it is reasonable to have this rule.

LEHRER: Well, look, you can't first rule out blockers on the ground, that the blocking predicates have to be nomological ones and then accept a predicate that is not a nomological predicate as a blocker. – So I think that you are putting too much weight on nomological predicates. Sometimes our reasoning rests on nomological rules, but sometimes it rests on very much more specific information like that all the ducks in Gerhard's Lake have a certain property.

SCHURZ: Yes, I mean, there is a difference between the second and first example. The second example is just not a disjunction with a single individual, but there is a reason, why ducks in lakes cannot fly. And even if this derived law contains an individual constant, it follows relevantly from some other laws and facts which are general laws. I don't think that a nomological predicate should never be allowed to contain an individual constant, I would not suggest that as a criteria for nomologicity. I would suggest more refined criteria. But whatever criteria I suggest, I want to say, that this framework is independent, because I am not forced to accept something as a rule. Look, the fact, that you don't want to admit the first example, but want to have the second example, is just manifested in the fact, that in your background system you don't have this rule, but you have the other rule. And one reason for accepting a rule is this that we believe that it contains some kind of causally relevant predicates.

LEHRER: I doubt very much, that the reference to causality is important. Take ravens. So if something is a raven, then it's black. But suppose there is a small number of ravens that are albino ravens and so they are white. And you have collected them all, you are fanatic of albino ravens. There might not be a causal explanation for why it is the case that if something is a Gerhard's raven, then it is white. This is a fact, I mean, it is a curious accident, that you happen to have this interest in collecting white ravens. But it does not follow from any causal truth.

SCHURZ: Well, the fact, that they are examples, where causality gives us not a criterion, does not mean, that the causality does not play any role. I do not claim that causality is the only or the dominant criterion, it is just one criterion. And I think, one should have a balanced view about causality, and obviously there are examples, where I don't know about causal relations, so I must have different criteria.

LEHRER: Well, I just think that some information that reflects accidental facts rather than laws may block inferences. So if that's the case, then I don't see, how you can complain that the accidental generalization that I have appealed to cannot be used.

SCHURZ: Well, it can be used, but it need not be used. I mean, if I think, that a rule is accidental, then I would not use such a rule. And if I think, that it is more than accidental, then I use this rule.

If something is a duck or is identical with Fido, which is my dog, then it can fly or whatever (laughing), – that is not a reasonable rule I think.

LEHRER: Almost all ducks can fly.

SCHURZ: Yes, let us agree on the following. We can distinguish two questions. The one question is, if you have a set of accepted rules, how to reason with these rules in a nonmonotonic way. All these rules are not deterministic, so how do we reason in a reasonable way with these rules, okay? And that can be represented in such a system. And if you consider belief change, if you consider all kinds of competing situations, then this system shows you, what is the right consequence to draw. The other question is, which rules we should accept. Well, this is a problem which can be separated from the first, because the question what really is a law and what is a merely accidental generalization, is a topic on its own.

LEHRER: Just one reminder. You have got to accept the role of accidental generalizations in reasoning. You cannot just say, that they don't have any role in reasoning.

SCHURZ: Yes, but obviously there are differences.

LEHRER: I agree.

SCHURZ: I mean, you know, if something is a derived law, it follows from a fundamental law and a fact, then although it contains an individual constant, it is law-like, which was pointed out by Hempel for instance. He said, that Galileo's laws about the earth contains an individual constant, but it is law-like, while the other examples which are about apples in the basket and so on, they are not law-like. So there are important differences. But this is the question, which rules should be accepted, I mean, this is not an objection against nonmonotonic reasoning. Nonmonotonic reasoning shows us how to deal in a computationally simple way with such non-deterministic rules without having probability values. Because your argument you can put forward against all kinds of rules it can be put forward against models of explanation; against every model which uses a notion, which hangs on the notion of law.

WEINGARTNER: The intention of nonmonotonic reasoning is usually to avoid certain redundances or also irrelevances in premises, and, not to allow, that any kind of inference keeps its validity independently of whatever you add to the premises, – I mean new knowledge or new information. Now, since this is the aim, it seems to me, that the framework you have presented, is still in some sense too weak, since it does not obey this aim, even if it is more detailed than what you have just written on the blackboard. Because in fact this is very rough: that oxygen is there or not – let's stick to this example – or the matches are wet or not wet. I mean, in fact we know very well, that a match will light even if it is a little bit wet or that it works also if there is very little oxygen. So if you express this in a bit more detail: O_1 to O_n say, or T_1 to T_n, then you can consider a replacement of a value O_1 by value O_2 and this by value O_3 and by value O_4 etc. And still the match will always light. But the possibility for replacement by n different values (with k parameters) shows, that this goes against what non-monotonic reasoning intended to avoid. Because it was intended to avoid that premises can be changed or replaced by other premises, without effect to the validity of the whole inference. So the conclusion of what I want to say is, that one probably has first to make this framework much more detailed to be realistically applicable and second to add restrictive rules in order just to avoid what one wants to avoid in nonmonotonic reasoning.

SCHURZ: Sure, I agree. I think, this is an important question you raised, which falls under the topic of verification and valididation of expert systems. They should not predict heavily errors. To make sure, under which conditions is it reasonable to accept such a law? There must be some kind of high probability. That is not a sufficient but the necessary condition.

ANONYMUS*: I found it interesting the details you have presented. I want to talk about the possibility to reduce these rules to statements, respectively to operations ending with material implication. I doubt the possibility to use material implication although I am well aware that it could be a good formal system. I will give you one of my arguments. I find it very interesting to speak about uncertainty and the respective reasoning with if-then-clauses without probabilistic values. But still, if it is possible to fly, then this is similar that to fly is not-blocked with respect to something. Isn't the fact, that something is not blocked with respect to something else a sort of a possibility clause and isn't it a sort of a modality? If it is, if this is a modality and could be accounted for in terms of a modal operator, then your rule "if A and not-blocked-A with respect to B, then B," could well be a sentence with the form of an implication. Then I maintain, that this implication, having this conjunctive component "possible to fly" or "not-blocked-A with respect to B", would in one way or another not be a material implication. So I maintain, that if you could reduce it to a system with a material implication, it is not adequate to the task of nonmonotonic logics. Or if you use another type of implication, another type of logic, it will be a sort of modal logic.

SCHURZ: There are several points involved in your question. The first point is, this "not-blocked operator", this was a modal operator in the original McDermott/Doyle formulation.The difference here is that in my system the consequence operation always leads to one fix point, while in the formulation where "possible B" means that the negation of B is not in the background system, this has not always one fix point. Except from that aspect, the idea is of course the same.

Second point. Indeed, in nonmonotonic logic there have been people who have constructed the whole nonmonotonic logic in modal logic. Brown is one man. He showed, that in S5, non-monotonic logic can be reconstructed. "Possible B" here is a modal operator. But as far as I know, the implication is a material implication. Third point. In nonmonotonic logic, "possible B" is a modal operator. In my reconstruction it has more content, because this not-blocked condition refers to other rules. It is certainly an intentional phrase.

ANONYMUS: Yes, sure, that's what my point was.

SCHURZ: If you want to have a semantics for this, then this must be some kind of intentional operator, certainly. But this "if then" (⇒) I guess, can be just represented by a material implication. Write in this way: Not-Blocked (A ⇒ B), then you have an intentional implication.

ANONYMUS: Why I maintain is that you could reduce this rule to a statement of an implication type. But this implication I do maintain cannot be the material one, if you have this intention operator in the antecedent. This is an argument of course, you could try to prove the theory you are talking about. I maintain, that it would not be the material one, or – if it is the material one, then the whole construction would not be adequate to do the purpose that you have set for, namely to get rid of this situation, where you have the detachment rule, and the possibility to infer B, no matter what new knowledge you have. I am quite sympathetic of the very idea to discuss changing the universe of discourse, changing the basis of knowledge. New knowledge is not necessarily compatible with the old one, some of the old knowledge has to be ruled out. I am fascinated by the possibility to approach this problem by means of various approaches, I mean, paraconsistently relevantly, nonmonotonically. I do of course respect this effort. I only maintain that if it is a material implication, then the initial task has not been fulfilled.

SCHURZ: I think, you may be right. What one should do is to try to develop a logic of this nonmonotonic implication. It's very interesting to think about the logic of this nonmonotonic conditions, because there are really problems. For instance think of contraposition. This rule seems to hold and then an interesting thing turns out. If you allow an irrelevant strengthening, then contraposition gives unreasonable, even false results.

ANONYMUS: My last question is more metaphysical. Taking S to be the sort of basic assumptions and trying to get the set of derived consequences. Could we dream of having in the set of derived consequences more information than in the set of basic assumptions? On what grounds, I mean, on what grounds?

SCHURZ: Yes, we have more information in the set of derived consequences than in the set of base information, because we have these nonmonotonic rules, which are based on inductive uniformities and we believe in them. I mean, the information in the derived set is not certain. It is more information and it is useful for acting, even if it is not certain.

KUTSCHERA: Are there any other questions?

LEHRER: Yes, it seems to me, that there are cases, in which there is an inference, that gets blocked. But then, what blocks the inference itself in a way gets blocked. I know that something is a raven, take that example. The question is, is it black? Well, it looks like you can infer from the fact, that it is a raven, that it is black. But now I have the additional bit of information, that this raven is in Gerhard's Lake and Gerhard collects white ravens. So that seems somehow block the inference from the information that I got to its being black. From the fact, that it is a raven, I am inclined to infer that it is black. From the fact, that it is a raven in Gerhard's Lake, I am inclined not to infer that it is black.

SCHURZ: But that it is white.

LEHRER: That it is white, perhaps. But the blockage can get on blocked in the following way. If I got the additional piece of information that this raven is in the Gerhard's Lake but was not one collected by Gerhard. – So that unblocks the blocker.

SCHURZ: Is it a problem?

LEHRER: I don't know if it is a problem. I am just asking, how you deal with it.

SCHURZ: Yes. – So, if it is a raven (writing on the blackboard) – unless blocked it is black – but you see I always read this in a way that a variable comes with it and it is quantified, you understand.

LEHRER: Right.

SCHURZ: Then, if it is a raven and it is in Gerhard's Lake, unless blocked then it is white. Well, and then I have: If it is a raven and it is in Gerhard's Lake and it flew into this lake, unless blocked, then it is black. Now you have the information that it is a raven, it is in Gerhard's Lake and it flew in. So the only rule which fires is the last one.

LEHRER: Yes. The only question I have is that you told us earlier that the second rule was not a proper rule because being in Gerhard's Lake earlier is not a nomological predicate.

SCHURZ: I think, it is a reasonable rule. To be in Gerhard's Lake, why not? I mean, there is a difference, whether I have a predicate like "$Cy \vee y = x$" or the predicate "to be collected by Gerhard". Why shouldn't it be, that the latter is reasonable but the former is not? I doubt, whether we can syntactically decide, whether a predicate is nomological. There are good reasons to think, that the question whether a predicate is a nomological, is dependent on background-knowledge.

LEHRER: I think, that if you admitted the rule with "Gerhard's Lake", a rule you are going to be admitting, that something, that contains that kind of reference to an individual like Gerhard, is nevertheless a nomological principle. I think you will have trouble in that way.

SCHURZ: Well, okay. I'll give you a criterion. It is just the idea of Hempel. A rule containing an individual constant is allowed as long as I can derive it from a fact and another law, which contains no individual constant in a relevant way. I can do it in this case, because you can say .

LEHRER: What's the law? The only principle, that supports the second rule is that Gerhard collects white ravens, and that's not a law.

SCHURZ: Well: For any person, who behaves rationally and who tells me this and this about his plans, will behave in this and this way. That's a very general law, provided the person is trustworthy. That's a non-monotonic law. If x is trustworthy (writing on the blackboard), and – let us speak about the case of promising – if x promises A, unless not blocked, then A will be the case. It is some law like this. I mean, you already used some laws like this, if I tell you I am collecting albino ravens and you conclude from that, in my lake there will be albino ravens. Isn't it?

LEHRER: It might just be, that I saw, that you were collecting albinos and I see you putting them into your lake. I don't know whether you will keep doing this, but so far you have been doing this.

SCHURZ: But there is a difference, if you are in the position where you have just seen me collecting ravens.

LEHRER: All I know is that this is in your lake and that you collect them. It is only a probability connection.

SCHURZ: But that I collect them is a statement about a disposition, not about a momentary act. So to infer that I collect albino ravens and there are white ravens in my lake, not by accident, but because I collect them, you need more information than just what you see at one moment.

LEHRER: I have been too long, I can't reply. Those are interesting answers.

KUTSCHERA: I have the impression that to employ conditionals or conditional probabilities is still the more general and, in the end, even the simpler way to deal with nonmonotonic reasoning. The system of your rules will become rather unwieldy in realistic applications. I find the idea of conditional logic attractive not to specify a list of relevant favorable or unfavorable conditions for the lightening of matches, but evaluate situations holistically. Lots of oxygen, after all, may make up for a little wetness.

SCHURZ: I don't understand, why you mean, that this is simpler. In conditional logic you mean possible worlds semantics.

KUTSCHERA: Yes.

SCHURZ: So – here you have a set of possible worlds. You have a preference relation between all those possible worlds. I am thinking about practical reasoning, not about what you can define in a reasonable way. There must be some kind of connection between probabilistic reasoning and this kind of nonmonotonic reasoning. It must be, what we call 'high enough' – the probability must be high enough. So if now change a parameter, which just slowly reduces the probability, then there must be some limit, where it is not high enough. It is not high enough for accepting this as a rule. This is certainly one criterion. I mean, the importance of nonmonotonic reasoning is to reason without probability values, but there must be some way of a probabilistic embedding of nonmonotonic reasoning. And this is interesting. Another problem is transitive chaining. If you have a lot of high probability rules it may be, that if you have a chain of rules, the resulting chain has not a very high probability, ceteris paribus other things. A nonmonotonic reasoning system should not lead us to heavy errors. This implies that we need a principle which restricts transitivity. It's an important topic in expert systems now to investigate and verify expert systems, such that they won't predict things which have low probability. And these are the topics which are connected with that. So one necessary condition is to require that the probability must be high enough and another condition is that if you have chainings of rules, the chains should not be too long. But all of these conditions are not sufficient as the points before have shown – there must be also conditions on the kind of predicates and so on. Now it seems to me, that it is computationally much more costly to calculate preference relations between possible worlds than just to use these simple rules, which tell me, in which situations which rules fire. If I were to try to write down a preference relation between possible worlds, it would be enormously complicated.

KUTSCHERA: It is the general idea that is much simpler. – You started out by telling us that the strategy of transforming "normic" laws into strict ones by spelling out the ceteris paribus condition into a detailed antecedent is not feasible, but I cannot see that your procedure is any simpler: Every known obstacle to the lightning of matches, and every known condition for these obstacles being ineffectual has to be accounted for in your rules. In realistic applications of your model the system of rules would be much more complex than in your simple examples.

SCHURZ: My examples are too simple?

KUTSCHERA: Yes, too simple.

SCHURZ: So on which reasons they are too simple?

KUTSCHERA: In the case of the matches you would have to consider still other conditions or rules.

SCHURZ: Aha, which?

KUTSCHERA: The match might be green, and green matches might mostly come from a firm whose matches don't light but break when struck. – You have also used a general preference relation for conditions. According to it your first choice, if the match does not light, is that it was not struck. If you know it was struck, the next choice is that it was wet, and so on. But the preferences may be different in different cases. It may, for instance, be more probable in some cases that the oxygen is missing than that the match is wet. Since the preference relation is to be determined by ceteris paribus rules, as it seems, your system of rules would have to vary from case to case, if not from match to match. You will have to state a rule for every circumstance that we consider relevant in the special case. But that would not work.

SCHURZ: No, you are right. But I don't see a counter-argument, I mean, I don't see any counter-argument in what you have found out. Of course we can formulate rules for every single situation. But these are non-deterministic rules. It may happen, that the antecedent condition is true, the rule is not blocked and the consequence does not follow, but only in rare cases. If that happens very often, I should better give up this rule, on inductive grounds. Of course there may be a thousand factors, which really play a role and I know only ten of them. But the point is, that this is the only way to deal with this situation, just because of what you said. Because there are so many factors, which are relevant, I can't really work with deterministic laws. So, what do we do then? I pick out of these – say hundred – factors, ten, which are really important. And with these ten factors I elaborate such a system. Practical reasoning does not intend to have an answer in every possible situation, but the point is, that I have a limited number of factors, none of them is deterministically sufficient, but I reason in a certain way of them, reflecting these kinds of preference relations. Of course if I would try to simulate a deterministic situation with this kind of rules, it would not make sense. It makes sense in a situation, where I have a limited number of important factors, and where there are clear non-deterministic relations between them.

KUTSCHERA: I understand what you mean, but I just wanted to point out, that the instrument of possible worlds or of probability theory is much stronger than what you have. It seems much more adeqate for dealing with these problems.

SCHURZ: Okay – I don't deny, that it is logically stronger, I am not ideological here, no, I like the translations. Probability theory alone will not be enough, it needs in addition maximized specifity conditions, like my "not-blocked" condition. Possible world semantics has preference relations instead. Okay, probability theory plus maximal specifity conditions, or possible worlds semantics plus preference relations is probably stronger than my rules. But the latter are much more apted for applied reasoning. Probability theory or possible world semantics cannot be applied to expert system reasoning for example. And it would be interesting to have some kind of relations, something, between these approaches.

LEHRER: So, your remark was about probability in expert systems and human beings. The fact, that it would not be plausible to build an expert system to compute the probability relations does not mean that the probability model is not a good model about us, because we don't have to compute those probabilities in order for them to explain our behaviour any more than the planets have to make computations in accordance to the law of gravity in order to determine their relations to each other. So, in other words, the explanation of human behaviour in terms of probabilities does not have to carry the burden of supposing that we in some sense compute the probabilities, any more than planets attracting each other or travelling on a certain orbit have to tally up computations. That is the way to answer your argument. It is right for expert systems, but the inference to the conclusion, that therefore the probabilistic model is not a correct model of us, just does not seem to me to follow.

SCHURZ: The point is maybe the difference between a totally connectionist model and a partly connectionistic model.

LEHRER: Right.

SCHURZ: If you think of the neuronal analogy of these propositions. In full probabilistic models then you have your statements, all of them are interconnected. If you want to compute a conditional probability, then all other facts, which you know, may be of importance. And if you want to say, they are not important, that again must be probabilistically calculated, because probabilistic independence just is defined as it is. So if you have – say (writing on the blackboard) A_1 to A_{10} factors and you want to know whether B or not, then you have to compute these conditional probabilities and these are very difficult. Now, in pure deterministic rules you have just one deterministic rule and if its antecedens is okay, then the consequence follows. So you have a purely local model here. But in my kind of model, you have because of these defeating conditions a partial connectionism. Some other facts are getting relevant, because of the non-blocking conditions, but not all. If I know, it's a raven and it is in Gerhard's garden, this is not enough, because that is a difficult situation, but if I only know, it's a raven, nothing more, I can conclude it is black – that's the advantage! In a full probabilistic model I would have to know the probabilistic relations to all other conditions. And now to the question concerning whether humans think fully connectionistically or only partially connectionistically – I don't know. Do you really think, that if I guess whether it will rain tomorrow, then there is some process in me which is calculating all these conditional probabilities without that I know it, and then the output comes? Or is it rather that I see a limited number of relevant facts and just handle to balance between them like here. Some may be defeating, but I *forget* the others. I don't know.

LEHRER: It is not obvious, that the connectionist model is true. So I will take the probabilistic model as a competitor to the connectionist model.

SCHURZ: I don't think, that it is a competitor. In Judea Pearl's work, the probabilistic approach is connectionistically represented.

LEHRER: Yes, that's possible.

SCHURZ: If you think of the propositions as the units and the probabilities just as the relations between the units, and then ask what happens, if you get here an initial probability on a piece of evidence – how get the probabilities propagated into the system? "Which changes occur?" That's his question and his model is very complicated. I mean, if you have a full probabilistic model indeed, you must consider all relations. And he is forced in his book then to reduce relations. He makes some simplicity assumptions, for otherwise all kind of propagations would be possible.

LEHRER: I thought that you thought that the computations required would be too many to account for on a connectionist model. So I was granting this assumption.

SCHURZ: No, sorry, the computation is in general too costly, but the only models which can do complex computations are totally connectionist models. Connectionist models can do very complex computations, because they compute step by step. That's this kind of computation by simulation. You just have to simulate the system and when the system converges, that's the solution! If it does not converge, it has no solution. That's the idea of connectionist models.

* This guest with interesting remarks participated in the discussion. Unfortunately the editor was unable to recover his name. Since not "names" but "contents" are important, the editor decided to document his remarks under the name "Anonymus".

ELENA KLEVAKINA-ULJANOV

THE CASE AGAINST VALUE-FREE BELIEF

The purpose of this paper is to consider the interrelation between belief and values. My claims are that (1) belief cannot be value-free, for it stands in extrinsic and intrinsic relations to values, (2) an attempt to reduce values attached to a certain belief to cognitive values would not be of great help for "purifying" the belief.

I.

Let us start with the question at what points and in what manner values enter into a belief. The answer is not as obvious as one might assume, because there is some kind of ambiguity in "belief". Thus the term can refer either to the content of belief, i.e. what is believed, or to the belief itself, i.e. to the statement as a whole (e.g., "a believes that p"). The latter is not really the content of belief, but a mental act of believing, i.e. something that a person "does". So the question arises: is it the believed proposition which is value-laden or the person's act of believing it?

My contention is that both the believed proposition and the believing itself stand in a certain relation to values. This relation can be either an extrinsic one when a value-component occurs in belief inessentially (that is, as an extrinsic component), or an intrinsic one when a value-component is embedded in a belief-statement. The evidence I want to present falls under three sections. In the first section I will focus on the extrinsic value-component inherent in the act of believing. In the second section I present some cases of the intrinsic relation between the content of belief and value-components attached to it. Finally, I will concentrate on some aspects of how valuations constitute reasons for believing. The question under discussion is *why* it is believed.

1.1

When I say that there is an extrinsic value-component in any act of believing, what I have in mind is a twofold relation in which the believing stands to knowledge or truth as a goal. Indeed one might think that knowledge is first of

P. Weingartner (ed.), Scientific and Religious Belief, 77–103

all a goal. Just consider, for instance, the first sentence of Aristotle's *Metaphysics*: "All men by nature desire to know".[1] Thus on the one hand, belief can be a means for achieving the goal or it can be immediately directed towards the goal. For example, a true belief can be a means for increasing knowledge or getting nearer to the truth. On the other hand, believing can lead away from the truth or knowledge as a goal, as false belief does.

My argument then runs as follows: the assertion of a statement such as "*a* believes that *p*" indicates that a person *a* thinks that by believing *p* rather than non-*p* he is closer to the truth, that is:

If *p* is true, then believing *p* is preferable to not believing *p*, or in symbols:

$$T(p) \rightarrow Pref(Bp/\neg Bp).$$

This search for truth is inherent to the act of believing and already at this point a cognitive value enters into the belief, even if extrinsically (i.e. inessentially). My further theses are: If *p* is true, then believing *p* is a value for all those who believe in *p*. In symbols:

$$T(p) \rightarrow \forall a(Val(aBp)).$$

Similarly:

$$Truthlike(p) \rightarrow \forall a(Val(aBp)).$$

Note, also, the cases of valuation with regards to religious belief, and in an analogous way also to ideology, although in ideology one has to replace 'Revealed' by 'Ideological' dogma. An essential point in religious belief is that it is connected with the last end (*EL*), so that if a person believes that some proposition *p* is revealed and he does not believe in *p* religiously, then he would not reach the last end. Or symbolically:

(1) $B((aB(Revealed\ (p))$ and $\neg aB_{R}p) \rightarrow \neg EL(a)$;
(2) $aB(Val(ELa))$;
(3) $aB(Revealed\ (p)) \rightarrow aB(Val(aB_{R}p))$.

For my current purposes further details are not important. The thrust of my argument is that the act of believing cannot be value-free since the goal of believing is truth, that is, a value.

1.2

Now let us turn to the question of how values enter into the content of belief. To clarify this point I will center the discussion around beliefs as basic principles of methodological norms, i.e. methodological beliefs.[2] Thus, the norm: "Base your hypotheses on all scientific information available", involves a certain belief that if a person violates this rule then he risks making a mistake, most likely because of a lack of information. But there are situations when all the information available in a certain field can hardly be interpreted consistently. In this case, the interpretation suggested could be dependent either on positive cognitive values (like observability, definability, truthlikeness, simplicity, etc.) or negative cognitive values (like non-observability or error). Indeed, beliefs which are intertwined with negative values are poor candidates for knowledge, for error as a kind of negative value is in conflict with the aim of science, i.e. to achieve truth.

As another example of the interdependence of norms, belief and values, consider the following normative principle of general methodology: "Confirm your hypotheses by seriously testing and criticising them". This norm presupposes a belief that a violation of this rule results in a number of false, yet confirming instances of the hypothesis. It is quite problematic, however, which tests should be considered to be severe and serious. In any case, this is the field open to valuation.

Taking all this into account, I am ready to conclude that since methodological norms are not value-free, the contents of their respective beliefs are also not value-free.

1.3

Further consideration concerning the value-ladenness of belief suggests a shift in the discussion from believing in general to the specific reasons for believing. In clarifying this point I will adhere first to the distinction between "merely explanatory reasons" and "justificatory reasons" as proposed by B. Winters.[3]

Merely explanatory reasons give an account of why something is the case. For example, "the reason for his believing that he would not win the lottery is that he had never been the winner in any game of this kind". Thus, the reasons under discussion do not justify a state of affairs, they merely offer an explanation of it.

On the contrary, the goal of justificatory reasons is to justify a belief. If so, this kind of reasoning is relevant to determining whether the belief is true or not – an observation which naturally suggests that one look for values in the belief which this reason supports.

For a better understanding of the connection between justificatory reasons and valuation a further distinction might be observed, that is, between cognitive justificatory reasons and non-cognitive ones:

(a) Cognitive justificatory reasons:

Suppose that b is an epistemic authority for a with regards to the field f. Then:

If p belongs to the field f, then a believes that b knows that p is the case. In symbols:

(1) $\forall p(p \in f \rightarrow aB(bKp))$.

Further, it seems to hold:

(2) $\forall p$(b claims $p \rightarrow p$ is consistent with what a believes).

Also, the person a may epistemically trust the person b. Then it holds:

(3) (a epistemically trusts b) $\leftrightarrow aB(\neg b$ lies to a).

Or, assume that a treats simplicity as an important criterion for believing the adequacy of a scientific hypothesis, then:

(4) $\forall p$(p satisfies the criterion of simplicity $\rightarrow aBp$).

(b) Non-cognitive justificatory reasons:

(5) b is a high value for a.

(6) $_a$(a axiologically trusts b) $\rightarrow aB(\neg b$ wishes bad things to a).

(6) $_b$(a axiologically trusts b) $\rightarrow aB$(b wishes good things to a).

(7) (a has a moral sense that "good" is an objective quality of (the object or the action) x) $\rightarrow aB$("good" is an objective quality of x).

As it is seen from the instances above, the characteristic feature of cognitive justificatory reasoning is that it basically involves cognitive values like epistemic authority, simplicity, etc., which influence a person's confidence that a belief-statement is true and, consequently, that it is worth believing. For example, as it is shown in (4), if p is a simple scientific hypothesis, and if a person a is inclined to change his values so as to put a greater emphasis on simplicity, then this change might enable a to believe p, i.e. to accept p as probably true.

Similarly, when directed by non-cognitive justificatory reasons, like a "high value", "axiological trust" or "moral sense", a person often is influenced by them in his belief decisions. As it is stated in (7), when a has to decide whether an object or an action x is good or bad, he makes a decision by means of his "moral sense", that is, by consulting his subjective feelings. Accordingly, his belief that "good" is a quality of x is rooted in value-laden reasoning based on his standards of differentiating between "good" and "bad".

Note also that these are not the only reasons that introduce values into believing. It is also the believing itself that brings about values in the reasoning. Thus in most (perhaps all) concrete cases of believing, in which reasons alone are not sufficient, a person has to make a volitive decision in order to justify his belief. To refuse to recognize volitive components in the believing is to leave an essential aspect of believing out of consideration.

In fact, with regards to believing I propose to distinguish implicit volitive components and explicit ones. Consider, for example, the case when a person knows that he does not know. What he has to do under such circumstances is only to believe. His believing then is based on volitive decision and in this sense it contains an implicit volitive component.

An explicit volitive component manifests itself when a person a strongly wants to believe that p:

Instances:

(a) $aBp \rightarrow a$ wants (p is true);

(b) $$aBp \rightarrow a \text{ wants } \begin{cases} \text{to consider } p, \\ \text{to think about } p, \\ \text{to find out about } p. \end{cases}$$

(c) $aBp \rightarrow a$ hopes (p is true); and also:
a hopes that $p \rightarrow a$ wants that p.

With the above instances in mind, consider, for example, L. Shestov's attack on what he calls a "rational philosophy".[4] For Shestov, this kind of philosophy from Aristotle on re-echoes, in some sense, the famous statement of Protagoras that man is the measure of all things. But what is our justification for believing that this statement is true? The answer of Shestov is that rationalism is the expression of our desire and of our will. "The root of all our philosophies lies, not in our objective observations, but in the demands of our own heart, in the subjective moral will ...".[5]

As another example of the valuation indicated above, it is plausible to ask why Priestley continued to believe in the phlogiston theory in spite of the fact that Lavoisier had already published his textbook. My point is that he wanted to believe it although his volition could have been guided by different reasons.

Perhaps Priestley did not want to switch to another theory because it required an enormous amount of institutional re-tooling, or perhaps, his conclusions concerning phlogiston were influenced mainly by heuristic values.[6] In any case, it was a volitive decision of Priestley to continue entertaining his belief even in the face of the emerging new reality.

Of course, the idea that human beliefs are dependent on volition to a certain extent is a very old one. One might recall Julius Caesar in this respect, who claimed that "[...] we readily believe what we wish to be true".[7] What is also of interest to me, however, is that quite often recognition of the volitive component is *the only way to explain* why a certain belief is sustained at present, that is to say, why it is now believed.

So my view comes to this: the process of belief acquisition intrinsically and extrinsically requires making value decisions. Consequently, human beliefs cannot be free of values – there is an ineradicable value-component in all belief-claims. However, there is still an interesting question remaining. Is a value-component in a belief-statement really ineradicable? Or, to put it in other words, can we reduce (i.e. decrease) or hope to reduce value-components in belief?

In what follows I want to show that any attempt to reduce the evaluative aspect in belief cannot be completely successful because there are always unresolved problems left. Nevertheless, with a remark of Professor Popper in mind, that "we can learn an immense amount even from unsuccessful or incomplete attempts at reduction, and problems left open in this way belong to our most valuable intellectual possessions [...]"[8], I will focus my attention on the problem of why we cannot succeed as reductionists with regards to the value-component in belief.[9]

II.

In my view, there are two possible ways of reducing value-components in belief – a "radical" one and a "modest" one.

2.1

What I call a radical way of reducing value-components is an attempt to neglect values altogether. Though a rather primitive kind of reduction, it is still well known from the history of science. Thus, for example, when constructing particular theories some philosophers choose to eliminate entities which they do not or cannot accept. As an example, consider the theory of matter elaborated by the extreme materialists (L. Buchner, K. Vogt, J. Moleschott), according to which the category of matter is treated as fundamental ultimate *sui generis*, whereas the category of non-material entities is simply eliminated, or in some sense, is reduced to the category of matter.

A similar question arises for our study: can we hope to reduce belief to a value-neutral concept. In fact, as Hume pointed out, "a wise man [...] proportions his belief to the evidence".[10] So what would be the outcome in an attempt to reduce "belief" to a notion of "convincing evidence" or "certainty"?

Clearly, after reducing belief to convincing evidence or certainty one could hold beliefs only on a limited range of topics. These could be, for instance, propositions of logic and mathematics, some cases of introspection and of simple perception. But if we were to widen the context of using belief-statements, the reduction would not go through. The problem is that it is impossible to say in general what counts as conclusive evidence, for this depends on the subject matter, on the circumstances and very much on the person who finds the evidence to be convincing for him. But if so, then we find ourselves faced with the phenomena of valuation once again. As a result, grounds which seem to be persuasive sometimes prove to be inadequate. Accordingly, a person can, in this sense, be justified in believing a proposition which is in fact false.

I will not offer any further observations concerning the consequences of reducing belief to other value-neutral concepts, although there are some possibilities similar to those described above. One more possibility, for example, is to reduce belief to "high confidence", that is, when there is some threshold degree of confidence r, so that a person believes only those propositions for which his degree of confidence is greater or not less than r. In any case, the thrust of my argument is that everything connected with "justification", "convincing evidence", "high confidence" and the like, is more or less a matter of valuation. As a result of this argument, I see no possibility for achieving fully satisfactory reduction with regards to belief-statements in which we would neglect values altogether.

2.2

Let us assume now that it is impossible to rid ourselves completely of values in belief contexts. Could we not then try to decrease the value-component in such contexts? The general way to do this would be to trace belief back to another concept which involves a less evaluative content. As a general example of the reduction of this kind, consider the Aristotelian theory of natural good, that is:

"An object (state of affairs) is a natural good (*NG*) if all persons desire it".[11]

Or symbolically:

$NG(x) \leftrightarrow \forall y(y \text{ desires } x)$.

In this case, the value-predicate "natural good" is reduced to a less evaluative one, "common desire", typical instances of which are "survival", "increasing knowledge according to abilities and interests", etc. Of course, one cannot say that "common desire" is completely free from values. Nevertheless, it is less evaluational also from the point of view of the fact that it can be investigated empirically. So in case of the Aristotelian theory the reduction in question results in some kind of a "bridge" (or a "bridge principle") between a value and a fact which can be confirmed empirically.

In my view, analogous reductions could be attempted with regards to belief. For instance, why not consider the possibility of reducing values attached to a certain belief to cognitive values, that is, to those connected with consistency, simplicity, truth, epistemic authority, etc. Although here values are not eliminated completely, still this could be a step forward to the value-neutrality of belief. What is also of interest here is how this "modest" kind of reduction would work.

Having in mind the distinction between the different aspects of belief proposed above, I will discuss three types of reduction in question, that is, with regards to (a) the content of belief, (b) the believing itself, and (c) reasons for believing.

$(a)_1$ With regards to the content of belief let us first apply the method of reduction to a scientific belief. Thus, when a scientist *S* believes that "a theory *A* is better than a theory *B*" and puts this into more specific words like "*A* is better unified than *B*" or "*A* is better confirmed than *B*", then the content of his belief, i.e. the statement, contains already a value predicate and consequently, is a value-statement.

With this fact in mind, one could try to reduce the statement "*A* is better unified than *B*" to less value-laden statements such as "*A* relates phenomena which were not related by *B* (but also those which *B* relates)" or "*A* explains everything from one unifying idea which is not present in *B*". Or, if we still consider the statement "*A* is a better theory than *B*" in a different meaning, then we could reduce it to "*A* is better confirmed than *B*" which in turn, could be reduced to "*A* has passed more severe tests than *B*", "*A* is nearer to the truth than *B*", etc.

At first sight it seems that we have reached our aim, that is, we have only cognitive value-components in the definiens such as "a unifying idea", "relating together", etc. The crucial point here is that the values in question cannot be treated as purely cognitive values. Rather, they are mixed with semantic values such as clear meaning and maximal truth, or methodological values such as testability and precision, even if these two categories partially coincide with what I have been calling cognitive values.

Therefore, with regards to the content of belief, shifting from values to cognitive values does not help too much, for cognitive values are intertwined with values of another kind.

$(a)_2$ For better confirmation of the thesis above, let us make a shift from scientific belief to a belief in teleological explanations. The essential point here is that in teleological explanations one explains facts which are not value-laden with the help of aims and goals which are certainly value-laden. Accordingly, there is a value-statement in the premise, whereas there is none in the conclusion. Thus

(1) If *x* believes that *p* is a high value (aim or goal) and *q* is a necessary condition for obtaining *p* , then *x* behaves in such a way that *q* obtains;
(2) *x* believes that *p* is a goal and *q* is a necessary condition for obtaining *p*.
(3) Therefore, *x* behaves in such a way that *q* obtains.

For example, Elizabeth I believed that not offending the official representatives of the Anglican and Catholic Church (if she pronounced her title unambiguously she thought she would offend them) is a goal and she believed that pronouncing her title ambiguously would be a necessary condition for reaching her goal. Therefore, Elizabeth I pronounced her title ambiguously.[12] Thus, what we have here is a reduction of a kind which has already been indicated – a reduction with a "bridge principle" like the Aristotelian definition of "natural good", i.e. from the state of affairs *p* as a value (goal, aim) to the state of affairs *p* as a desire, wish, etc. Observe, also, that in teleological arguments the fact explained does not, or at least not apparently, contain values, though they may be hidden. To be more precise, here value-statements occur as explanatory statements of a conclusion which does not contain a value-statement:

"*E* uses an ambiguous language". Thus the reduction goes from the premises to the conclusion. However, in the explanation of the conclusion there are still values.

(b) My argument so far has been directed to one point – whether it is possible to diminish the value-component in the content of belief. However, in discussing reduction with regards to the believing itself I should like to reflect on one more point, that is, the consequences of the attempt to reduce.

As a starting point for a discussion I appeal to a principle of epistemic preferability developed by R. Chisholm.[13] Analyzing the epistemic status of a given proposition *h*, for a person *S* at *t*, Chisholm wants to know what is preferable epistemically: believing *h*, disbelieving *h* (i.e. believing the negation of *h*) or withholding *h* (i.e. neither believing, nor disbelieving *h*). To summarize the relevant possibilities, he uses the following abbreviations: '*Bh*' for 'believing *h*', '*B*¬*h*' for 'disbelieving *h*', '*Wh*' for 'withholding *h*', and '*P*' for 'epistemically preferable to'.

Without attempting to reproduce precisely Chisholm's assumptions (which he takes as axioms), I will note certain principles which help to exhibit the relations among the concepts stated above:

If *h* is true then *Pref*(*Bh*/¬*Bh*), where ¬*Bh* means not-believing *h*;
If *h* is true then *Pref*(*Bh*/*Wh*);
If *h* is true then *Pref*(*Bh*/*B*¬*h*).

Chisholm deliberately reduces "*h* is better then *y*" to "*h* is preferable to *y*". For, according to Chisholm, "*h* is preferable to *y*" need not have implications with respect to the goodness or badness of *h* or *y*, and "*h* is epistemically preferable to *y*" need not have any implication as to the reasonableness or unreasonableness of "*h*" or "*y*".[14] With regards to epistemic preferability, Chisholm suggests, it is much more important that the relation *P* is transitive and asymmetric.

Note, however, that Chisholm's principle of epistemic preferability is powerful when limited to belief-relations in epistemic logic, but it is inefficient when applied to the belief itself. Regrettably, the consequence of the reduction discussed above is that it gives no room for our investigation into motivations. Accordingly, it leaves behind the so-called "why-question", i.e. why withholding *h* is more reasonable than believing *h*, and consequently, why believing *h* is more reasonable than disbelieving *h*. So as a next task, I will analyze the relation of motivation to belief with respect to the consequences of reducing valuation. That is, I now turn from the believing itself to reasons for believing.

(c) In general, one could discuss the consequences of the reduction with regards to two kinds of values, namely, positive cognitive values and negative cognitive ones. In this paper I will limit myself to positive cognitive values only. For in either case, my point is that the degree of valuation in belief cannot be reduced to cognitive values exclusively.

I will illustrate this point by adhering to the notion of inconsistency and simplicity. The problem under discussion is whether these values are by themselves sufficient to determine the belief of a scientist when he prefers one theory to the other. Following T. Kuhn, I will consider how consistency and simplicity (as evaluative criteria) functioned in the choice between the heliocentric and geocentric systems.[15]

Kuhn maintains that, as astronomical theories, both Ptolemy's and Copernicus's were internally consistent, but their relation to related theories in other fields was rather different. As is known, the central idea behind the received theory was that the earth is stationary at the center of the universe. Accordingly, the doctrine explained among other things how stones fall, how water-pumps function, etc. As to the heliocentric astronomy, and its commitment to the motion of the earth, it was inconsistent with the existing scientific explanation of these and other terrestrial phenomena. Thus, the consistency criterion supported the geocentric tradition.

Simplicity, however, favoured Copernicus, but only if it is evaluated in a special way. On the one hand, both systems proved to be basically equivalent when compared from the point of view of the computational labour which is necessary for predicting the position of a planet at a particular time. The system of Copernicus did not offer astronomers any labour-saving technique and it was not simpler than the system of Ptolemy in this respect. On the other hand, Copernicus's system was simpler with regards to the amount of mathematical apparatus required to explain the essential qualitative features of the planets (limited elongation, retrograde motion, and the like) – Copernicus required only one circle per planet, whereas Ptolemy required two. In this sense, the Copernican theory was simpler and this is the fact which helped to influence the choices made by Kepler and Galileo and it is also contributed to the victory of Copernicanism. But what is noteworthy here is that the sense of simplicity was not the only value available nor even the one most natural to professional astronomers whose task was the actual computation of planetary positions.

Indeed, difficulties in applying standard criteria of choice are typical when scientists must choose between competing theories, and two men having the same criteria for choice may reach different conclusions. For example, scientists may interpret simplicity differently or they may have different opinions concerning the scope of the fields within which the consistency criterion must be met. In any case, even when two theories are compared via cognitive values only, there still remains a gap which is filled by an evaluation of the scientist which is not cognitive. Rather, it depends on the goals which a particular scientist has (note, for instance, the case with Priestley), on his wish to achieve this goal, on his specific world outlook, which can be determined by a cluster of cognitive, moral, political or even national values.

To conclude, the foregoing discussion of the connection between belief and values has been largely of a negative character, being concerned primarily with the problems and difficulties that confront any attempt at reducing the value-component in belief. However, certain positive conclusions do emerge.

First of all, one has to meet the fact that belief is intertwined with values which, in the extended sense, consist in or arise from goals towards which we strive. From this point of view, it is plausible to differentiate the primary type of values from the secondary. Primary values come with the direct goal, i.e. to achieve knowledge or truth. Secondary values are the by-products of intermediate goals which are encountered on the way to achieving the primary one. Rationality, for instance, is a secondary value from this point of view, if we treat it as a method which increases the effectiveness of gaining the primary goal. The outcome suggests that believing extrinsically requires making value-decisions. And if, as I have stressed above, we take intrinsic values into account, the evident conclusion is that believing can never be value-free.

But the question remains: can a value-laden belief be objective? I *believe* that it can. And the first step towards establishing its objectivity might be to stress a greater reflective self-awareness to which value-judgements might accompany a belief, or even, what value-decisions have to be made in each particular case of believing.

NOTES

1 Aristotle, Metaphysics, p. 1552.

2 See P. Weingartner (1980), pp. 218–219.

3 See B. Winters (1980), p. 6.

4 For Shestov's criticism of rationalist philosophy with regards to wishful thinking see, especially, Shestov (1977).

5 Op. cit., 60 (Part I, section 118).

6 Noretta Koertge advocates these motives of Priestley when she gives her account of Priestley's assessment of the costs of being wrong in believing in the phlogiston theory long after the publication of the results of Lavoisier. For more details see N. Koertge (1985), 185.

7 Quoted in W. Stark (1958), 47.

8 K. Popper (1982), 133.

9 For more details on the problem of reduction in epistemic matters see my 'On the Problem of Reducing Value-Components in Epistemic Concepts' (to appear).

10 D. Hume (1975), 110.

11 Aristotle, Nicomachean Ethics, 1729–1867.

12 See E. Nagel (1971), 552–553.

13 See Chisholm (1970), 294–301.

14 Op. cit.,294.

15 See T. Kuhn (1979), 323–325.

REFERENCES

Aristotle: 1985, 'Metaphysics', in *The Complete Works of Aristotle*, J. Barnes (ed.), Princeton University Press, Princeton/New Jersey.

Aristotle: 1985, 'Nicomachean Ethics', in *The Complete Works of Aristotle*, J. Barnes (ed.), Princeton University Press, Princeton/New Jersey.

Chisholm, R., 'On a Principle of Epistemic Preferability', *Philosophy and Phenomenological Research* v.XXX (September 1969–June 1970), 294–301.

Hume, D.: 1975, *Inquiries Concerning Human Understanding and Concerning the Principles of Morals*, Clarendon Press, Oxford.

Klevakina, E.: 1991, 'On the Problem of Reducing Value-Components in Epistemic Concepts', in *The Problem of Reductionism in Science*, E. Agazzi (ed.), Kluwer Academic Publishers, Dordrecht/Boston/London.

Koertge, N.: 1985, 'On Explaining Beliefs', *Erkenntnis* **22**, 175–186.

Kuhn, T.: 1979, *The Essential Tension*, The University of Chicago Press, Chicago.

Nagel, E.: 1971, *The Structure of Science*, Routledge and Kegan Paul, London.

Popper, K.: 1982, *The Open Universe*, Rowman and Littlefield, New Jersey.

Shestov, L.: 1977, *All Things are Possible*, Ohio University Press, Athens.

Stark, W.: 1958, *Sociology of Knowledge*, Routledge and Kegan Paul, London.

Weingartner, P.: 1980, 'Normative Characteristics of Scientific Activity', in R. Hilpinen (ed.), *Rationality in Science*, Reidel Publishing Company, Dordrecht, 209–230.

Winters, B.: 1980, 'Reasonable Believing', *Dialectica* **34**, 3–15.

DISCUSSION

KUTSCHERA: It's only a small point that I would like to make. If you take values in the objective sense, your principle "If *p* is true, then it is valuable that *p* is believed" is plausible in many cases (it might not be valuable, though, that *p* is believed for the wrong reasons, or that an assassin truly believes that his victim is defenseless). But if you take values in the subjective sense this wouldn't be so, for the subject might believe that *p* is wrong. So what you should say is: if somebody believes that *p* is right, then it is a value for him, that he believes, that *p* is right.

LEHRER: I see you are inclined to distinguish an objective and a subjective sense of value. Your principle then catches an objective sense of value like in utilitarianism. Thus, there is an objective sense of right, if an action maximizes happiness. And there is a subjective sense of right, if you believe, that it maximizes happiness. But it just seems to me that both notions are appropriate to be distinguished.

KUTSCHERA: Yes, and it is the objective notion that seems to be intended here.

KLEVAKINA-ULJANOV: Yes, I agree, although with some specifications. I accept the general distinction between objective and subjective values. But I am in a doubt whether this distinction is profitable when applied to values which enter into a belief. My doubt depends partially on the definition of the concepts of axiological objectivity and subjectivity. Thus, if we proceed from the assumption that values are axiologically objective if they are independent of desire, or if we follow the approach of Prof. Bunge that values are axiologically objective if they are rooted in basic needs or legitimate wants, then my question is: could there be an exclusively "objective" belief, i.e. which is entirely determined by basic needs and legitimate wants without subjective factors coming into play? I think, even if it is true for some beliefs, it is not the case to be generalized. And this is primarily because of the relation between belief and choice, in my opinion. Thus, in coming to believe *p*, we come to favor *p* among alternatives *q*, *r*, *s*, etc. So we make a volitive decision to choose *p* in spite of other possibilities. Sure, choice is not equivalent to desire – the former extends more widely and therefore, it is not completely subjective. However, bearing the remark of Aristotle in mind that "choice is a deliberate desire of things", I would say that it never can be exclusively an objective matter. For instance, when we choose to believe *p* because it is supported by conclusive evidence (i.e. the evidence is not merely good but is as good as it can be, that further investigation has lost its point), our choice is still not exclusively "objective", for even in this case we are guided by desire, that is, a desire to have a true belief. Viewed in this way, choice introduces a volitional component into the process of belief acquisition. As a result, the latter can't be properly construable in terms of objective values only. At least, we are confronted with a process where objective and subjective values are intertwined so that the distinction between objective and subjective values with respect to a belief makes little heuristic sense to me.

WEINGARTNER: I have a comment to your remark on Aristotle. He had a bridge principle with which he connected natural objective value (good) with desire:

A state (of affairs) p is (naturally) good if and only if all men desire p.

Thus the right side though referring to men, does not refer to the desire of one or other particular man or to a group or society (i.e. subjective) but to all (i.e. objective in this sense). Examples are according to him: All men desire to know (Metaphysics, first sentence), all men desire to survive, all men desire to live in a society ... etc. It is important to notice that such natural goods which are objective in the above sense are not claimed (by Aristotle) to be highest goods and can be violated (think of suicide). Though they are natural and very basic, we might better call them basic needs. The highest good according to him is happiness of the individual person in a society.

The right side of the above bridge principle is clearly empirically testable in some way. The whole equivalence is testable only if also the left part is testable in some way (independently of a criterion of desiring). But this is exactly the difficult question. Let us consider a statement which has the same logical structure for comparison:

A thing is cheap if and only if all men buy it. First there are such things which are bought by all men, thus like above we have examples, for instance: ground food. But is it always cheap (relative to some average income in a country)? Probably not always. Here we can test the right part and the left part seperately. And it turns out that the statement is not always true, but perhaps in the majority of cases.

Coming back to Aristole's bridge principle: One thing is clear: If I accept the bridge principle as an axiom or as a true definition, then what is a natural good can be tested empirically. Or more modestly: If I accept the bridge principle as some reasonable interpretation (or explanation) of "natural good", then by the empirical testability of the right side I have some kind of empirical access to that concept. Applied to belief then I'd like to say that in cases the values involved in belief are those basic needs of mankind, we have a more objective evaluation entering the belief which is also closer to empirical access than in cases the values are restricted to social groups or individuals.

KLEVAKINA-ULJANOV: Thank you for this important addition. I am with you that one can achieve a complete reduction, if one part of a proposition can be put to test independent of the other, and a half-reduction when the left part of a statement cannot be confirmed irrespective of the right one, as it holds for a "bridge" statement, so to speak. The former is thus in this very special and limited sense "reducible" to evidence, whereas the latter indicates the lack of direct access to the empirical reality. This is a good point in which I even find a reinforcement for the idea that with respect to a belief claim one can get only a half-reduction, at least as far as G-beliefs (the notion introduced by Prof. Weingartner) are concerned. My contention is that like a "bridge" statement, G-belief embodies valuation, conveying reasons, premises or assumptions rather than giving evidence. Accordingly, justifying G-belief is a matter of estimating reasons or grounds, that is, of what is not a fact but a mere premise or a conjecture. The case is even complicated by the fact that the background for some beliefs could be emotional rather than cognitive reasons. Here the example of Prof. Lehrer comes to my mind that a husband's belief that his wife is faithful to him might give him needed emotional security and his belief might be held for this reason, rather than because of a concern for knowing the truth. This or that way but a complete reduction is unattainable with respect to G-belief, in my view. How far the reduction would go through is a matter of a degree of valuation. Thus, one can have a degree of valuation very low when G-belief, having a purely cognitive role, is based on good evidence, i.e. convincing enough to produce a reliable belief. The degree of valuation can be rather high, however, when a person's evidential position is not strong enough to justify the proposition and, what is more, the belief was formed for emotional reasons, as it is described in the example above. In all other cases (for instance, when it is urged that God will reward those who believe in Him – the assurance that beliefs of certain kinds may be rewarded) one can have the degree of valuation which would remain in-between the summit points in question. Having all this in mind, I would say that both "bridge" statements and belief-claims are similar in so far as the problem of justification is concerned. And as we cannot achieve a complete reduction in a "bridge" case, there is no point to strive for it in belief-claims, I think.

LEHRER: I want to raise a couple of objections. I am not much committed to them, but they are the standard objections to the kind of account you are offering. The first one goes like this: There are beliefs that we have which we cannot help but have. So having them really does not depend on values, like e.g. all of us believe, that there are people in this room. – It doesn't matter, what the values are. Here is something that you believe immediately without values coming into play. So isn't that a case in which you have a belief without valuation entering in?

KLEVAKINA-ULJANOV: Yes, I think, the case you have presented has no direct connection with valuation, at least, in the sense I use the term "to evaluate". What I have in mind is a particular kind of valuating, i.e. one that includes reflection and comparison. In my account, then, "to evaluate" means "to appraise", "to estimate". Coming back to your example, I would say that it deals with a sort of simple perceptual beliefs which are mostly acquired automatically, *without valuating*. I mean it is an automatic process that my perceptual experience of a physical object or the room I sit in, results in a certain belief. Analogously, there is no point to look for values in beliefs about one's psychological properties (e.g., "being cheerful"), as these properties are self-presenting and so the corresponding belief comes "immediately", without reflection. In any case, the belief you have exemplified seems to be value-neutral: I come to believe that there are people in this room because my perceptual organs are working and because it is so. In fact, being in a position to claim such a belief is equal to claim knowledge, I suppose. But my purpose was not to talk on this. I was concentrated on G-beliefs which stand in opposition to knowledge, that is, if a G-believes that p then a does not know that p.

LEHRER: Okay, I see your point is that there are certain propositional attitudes, which may be free of evaluation, but that's not the one you are concerned with.

KLEVAKINA-ULJANOV: You are right. Again, my concern is a belief which is arrived at when, *after estimating state of affairs,* one finds the evidence sufficient for its support, though the fact that he finds the evidence sufficient is not a proof that it is really sufficient like it is in a claim: "These manuscripts are believed to date from the XIth century".

LEHRER: Well, but let me raise, anyway, the standard reduction of justification for a belief even in the sense you have in mind, I think, which is the reduction to probability. So justification is reduced to probability and how probable something is, say how probable it is on the evidence, according to such a view, does not involve evaluation. It is just a matter of probability or truth frequency. So what would you say to that attempted line of reduction? You said something about this in terms of confidence level, but it was not clear to me why you would think that a confidence level or probability level would necessarily bring value back in again.

KLEVAKINA-ULJANOV: Well, when I spoke about confidence I did not mean exactly probability.

LEHRER: I see, okay. But still what would be your objection to the claim that you can reduce justification to probability and if you reduce justification to probability, you will eliminate value.

KLEVAKINA-ULJANOV: I think the answer to this question will rather depend on the context. Let me elaborate my point a bit. First of all, I would not deny that there could be a case when a probabilistic kind of belief is value-neutral in the sense that a believer need not make a volitive decision by comparing facts. Consider, for instance, a believed proposition which is resultant from statistic probability: "The probability is .91 that seventeen-year residents of Russia have finished a secondary school". I guess, this proposition does not involve special implications concerning one's evidence for believing. In fact, this probability statement is a translation of a claim: "Ninety-one percent of the seventeen-year residents of Russia have finished a secondary school" and so, one should not "weigh" pro and contra to acquire a corresponding belief. It is a statistics, we say, and like in case with sense-data, the corresponding belief need not to involve evaluation. Of course, you may share the idea expressed in the joke: "There is a mere lie, there is a foul lie, and there is statistics", so that you are not inclined to believe statistical claims at all. But then, I would say, it is your subjective presupposition which should not be generalized, just as we don't assume as a basis a position of a sceptic who does not believe the results of perceptual experience.

Or, let's take inductive probability statement, as an example. As it is suggested by the logic of inductive probability, when the logical or syntactical properties of any statements, p, q, r are ascertained, it is possible to determine whether q is probable in relation to p and whether q is more probable than r in relation to p. If so, the term "probable", when used inductively, can be regarded as a purely logical concept. Accordingly, a probability statement which says that, in relation to p, q is probable, or more probable than r is a purely logical statement.

In my view, valuation comes to light with a man's *decision* to appeal to probability to show that he has adequate evidence for his belief. In this case, his believing p (which he assigns a high probability of being true) is determined by his goal, i.e. to get truth, having made the choice to favor p he is able to satisfy his desire, to have a true belief. Viewed in this way, probabilities for believing or disbelieving *p* are a function of values, namely, truth-seeking and desire and in this sense values are not escaped, I suppose, even though valuation enters into a belief inessentially, i.e. as an extrinsic component. Perhaps we could find even an intrinsic value-component in a belief of the probabilistic kind, that is, when we have no certainty concerning *p* but only some inducements to believe that *p* is true. It is evaluation, then, which would determine our choice to believe *p* and not to give preference to plausible alternatives.

HICK: Let me note that you might have another line of response to Keith Lehrer. You might say that in the case of sense perception, where we have an immediate, spontaneous belief that there are other people in the room, we are biologically programmed to believe that things are as they present themselves to us. But this biological programming presupposes or is directed towards some value, – survival, for instance, – which enables us to operate in the world as it is. And then again, in the case of probability, if you reduce belief to choosing the superior probability you might ask, why should we prefer the more probable to the less probable; and again the value of truth might be assumed.

LEHRER: You are going to say in assigning that high probability just so far as to produce a belief. You then have to answer the question of why I should care about high probability. The answer seems to be that the reason you care about high probability is that you think it is a guide to truth and, that is, to value.

HICK: Yes, exactly.

KLEVAKINA-ULJANOV: Thank you very much for the answer. I agree with you that we could find values even in simple perceptual beliefs, if to consider valuation in a wider sense (i.e. to cover in addition to "good", "desirable" or "worthwhile" all kinds of rightness, obligation, virty, beauty, holiness, etc.). But I still think that if we use "valuating" as synonymous to "estimating" or "appraising" I would not attribute evaluation to a simple perceptual belief.

KUTSCHERA: Do I understand you correctly, that, in case you have, let's say, four pieces of evidence relevant for a proposition *p* and all four speak for the proposition, then in effect you have no choice but to believe it, and if you have no choice, no values enter. If only three pieces of the evidence speak for *p* and one speaks against *p*, then there is, roughly speaking probability of 3 to 4. To assign to *p* this probability you still have to make no choice and no values enter. But if you go on from this probability assignment to a belief in the strong sense, a conviction that *p* is the case, then there is a decision, and then values enter. Would that be your position?

KLEVAKINA-ULJANOV: Not exactly. Certainly, that's a crucial point, in my view: values enter when there is a choice. That is, when a subject S has no adequate evidence for a proposition *p*, he has to determine whether *p* is more probable than not (in the sense whether it is evident or not) and that is the point where values intervene. But I also think that if a proposition is probable in relation to "total evidence" available of any subject S or, to put it in your terms, when all pieces of evidence speak for the proposition, the corresponding belief is still evaluated, noting that it includes an extrinsic value in the sense described above. However, I am not inclined to ascribe evaluation to beliefs which are acquired automatically, without "weighing" evidence for its support. Of course, we may say of any statement that it contains search for truth as an extrinsic value, but as I said earlier, it is not the point of my concern, for it does not give too much for the apprehension of the origin of a belief.

WEINGARTNER: Truth you could always have as an extrinsic value additionally. There is certainly a big difference between what you called extrinsic and intrinsic values in respect to beliefs. In this respect it is very important if there is some component of the will for the decision. These are of course much stronger components in respect to valuation. But the very weak point in respect to valuation that beliefs always have some extrinsic relation to some aim, say truth, this I think, is in a good sense independent of the other intrinsic evaluations. Although I agree with you in the sense that it would be rather pathetic to underline this so much, since the intrinsic values are more specific, more interesting and bring up more details of belief than the extrinsic ones.

SCHURZ: I think, one could put the question in the following way: Is it true for a certain class of beliefs that values are always among the reasons for these beliefs? If I have this question in mind, then we should formulate it in a certain way so that not to receive trivial answers or let's say, it would be a nontrivial question. In some sense it has a trivial answer, I think. For, how do we distinguish reasons from causes? We distinguish it by saying, that if something is a reason, a reason for an action, there must be some goal involved in a way, that's just how we distinguish reasons from causes. So, whenever we say, (believing) A is a reason for believing B, of course there must be some goal among these reasons, in most cases it is truth. I mean, why do we believe, that probability is an equivalent for belief? Because we think, that probability leads us to truth and we want truth. So in this sense the answer must be, I think, always yes. Reasons at some level of justification imply some goals, cognitive and internal goals, too. But I think, there should be a non-trivial version of this question. And it seems to me that what you had in mind was a stronger claim. Maybe you could make your point more precise?

KLEVAKINA-ULJANOV: Thank you for your remark. Of course, the fact that reasons for believing involve goals and hence, values is not a point at issue. But my question is *to what extent* they are evaluated so as one could speak about objectivity of a corresponding belief. The problem is that in believing, it is difficult to draw a line between reasons (which are goal-seeking, mostly even truth-seeking, by nature) and non-reasons or psychological causes (particular associations, memories, habits, traditions), that is, the distinction you've just mentioned. For believing is a psychological event, no less than wishing. And so, with respect to a belief, having a reason is itself a kind of a cause. What I have in mind is that our system of belief acquisition is less than exclusively "reasonable", i.e. based on reasons. Mostly, our decision to believe is the outcome of a certain "seeing" reality in the sense of Wittgensteinian concept of "seeing as" or "paradigm" of T. Kuhn. This kind of "seeing", which constitutes the background of our beliefs, could be supported by different cause-factors, even by a non-reasonable system of rules. For instance, this is the case with an experienced driver who believes the adequacy of traffic rules, although the traffic system does not provide reasons for these rules or for the system itself. So, in the last analysis, we could come to Darwin's and Freud's contentions (though seemingly incompatible ones) with Darwin's stress on a "rule of chance" and Freud's accents on a subconscious will: what superficially seems to be a "well-founded" belief, when analysed, may become a by-product of a non-reasonable system of rules, applied by chance and regulated by subconscious goals. Anyway, what one can believe is normally determined not only by reasons but also by non-reasons – which one has – the fact which calls the objectivity of any belief in question.

SCHURZ: Yes, that's an interesting version of this question, if you draw a distinction between internal, say truth-internal values, i.e. truth and those following from truth (like truthlikeness and so on) and external values which are non-reasons.

KLEVAKINA-ULJANOV: Yes, but in my view, this distinction only presses the case against value-free belief. Being a psychological state, the believing always involves internal and external values. These two kinds of values vary independently. That is, sometimes one predominates, sometimes the other. But it's only a matter of degree. And though marked by a difference in degree of valuation, the believing is still a subjective enterprise.

KUTSCHERA: One more question. If we just assign probabilities there seems to be little room for a value dependence. Our decisions depend on valuations, certainly, but these are preferences for aims and not about our epistemic states. But what about cases where you cannot be content with probabilities, but have to base your action on a firm conviction. William James has a name for such situations, I don't know how he called them, 'existential situations', or something like that.

HICK: 'Forced options', was his phrase, I think.

KUTSCHERA: 'Forced options', yes. There are other, less dramatic cases, too. For instance a judge or a referee often has to rule one way or the other in the face of insufficient evidence. Could you say something about such cases, how epistemic values enter on the scene and how epistemic decisions are guided by them?

KLEVAKINA-ULJANOV: To explain, why the step from probabilities to convictions is necessary, I would, first of all, make a distinction between mathematical probability and probability in a general sense or what Cournot calls philosophical probability. In former case the grounds of our preference are subjected to purely logical or mathematical demonstration, whereas in case of the latter we base our choice on variable, subjective probability. For example, when a scientist chooses between hypotheses, he refers to certain criteria, such as simplicity or consistency which support his conviction in the adequacy of the decision. But this feeling of certainty does not change the fact that what the scientist finds to be more probable depends on a limited extent of his present knowledge and thus on a variable factor. Moreover, to justify the choice, the scientist has to introduce other criteria which are mostly not amenable to purely mathematical formulation. And here is the problem: we cannot indeed stay with mathematical probabilities alone, rather, we are doomed to deal with philosophical probabilities in the majority of cases. This is, therefore, the reason *why* we have to make this step from probabilities to convictions, in my view. To answer the question *when* this step is necessary, I would say like W. James: whenever one must make a genuine option (i.e. an option which is of the forced, living and momentous kind), one has to "slide" from probabilities to convictions. For, the nature of a genuine option is such that it cannot be decided on purely intellectual grounds. Being mostly based on the insufficient evidence, it always involves a volitive decision to acknowledge a proposition as being true. So, when we deal with a philosophical probability, moreover, we are faced with a genuine option, then it is the matter of our subjective conviction what we judge to be more probable than not.

LEHRER: I have a rather long remark, that I would like to make, but it is not very urgent.

HICK: Mine is a sort of appendix, so please proceed.

LEHRER. Okay, this is just an alternative account, that supports your view and I simply wonder about your opinion of my argumentation. You probably know, that there is an account that says that belief and, particularly, rational justified belief, results from maximizing expected epistemic value. So, if we take the proposition *p*, the expected value of accepting *p* or to put it this way, the degree of justification for accepting *p* (or the degree of rationality in accepting *p*) is based on a consideration of two alternatives: either you believe *p* and you are right or you believe *p* and you are wrong. There are always these two alternatives and the question, then, of how justified you are, is a function of how much value you attached to being right and how much value you attached to being wrong. And I guess that this expected value or the degree of justification (or degree of rationality) is an equivalent to the value that you attach to accepting *p* or, let's use your term, believing *p*. So if you look in this way, then the degree of justification or the degree of rationality is a degree of the expected epistemological utility and, therefore, it is a function of two values, that is, the value you attach to truth and the disvalue you attach to error. That idea seems to be pretty sympathetic to the kind of view that you are putting forward.

KLEVAKINA-ULJANOV: Thank you for the interesting addition. I would also say that epistemic utility, namely cost of errors, is an important determinant of believing, as well as probabilities or weights of evidence. And, in some sense, the degree of justification or even better, the degree of rationality of a belief is rather determined by pragmatic considerations, that is, it varies in accordance with the seriousness of a projected mistake. It is a fact, for instance, that there are two different standards of proof in the British law. In criminal cases members of the jury are instructed in such a way that they should be absolutely (i.e. "beyond reasonable doubt") sure of their verdict, for if they are mistaken an innocent person could be put into prison. However, in civil cases they only should be convinced "on the balance of probabilities" that a person is guilty. This is a good confirmation of your case, I think, that the degree of justification of a belief is dependent on the value one attaches to truth and on the disvalue one attaches to error.

HICK: Let me come to my appendix. This is not to do with your epistemology. It just concerns your passing reference to religious belief, your thought that people might believe that their future and their salvation depends on believing rightly. I just wanted to mention, as illustration of the diversity of the trends in religion, a very interesting case of the Buddha. Someone came to him and listed a series of propositions, such as the world is infinite in space, the world is not infinite in space, the world is eternal, the world is not eternal. Another was a very contemporary one, namely, whether soul and body are identical. And they asked him, 'Tell us, what is the truth of these matters?' But he refused to answer, saying, 'You do not need to know the answer to these questions in order to attain enlightenment'. And indeed if you make it your great object to find out – if you believe, as we would say in Western terms, that salvation depends on right belief, – this will get in the way of salvation. So this is opposite to the view that in order to achieve salvation it is essential to believe rightly. According to the Buddha it is totally unimportant.

KLEVAKINA-ULJANOV: Yes, but as a matter of fact, I did not intend to treat the longing for salvation as an essential point in all cases of religious believing. No doubt, as any belief, it may be caused by different factors. But would you not think that the salvation argument which I have brought about, is also quite possible and if so, it supports the idea that there is a strong volitional element in a religious belief. Anyway, all I want to stress is that reasons for believing religiously involve volition.

HICK: Yes, okay.

LEHRER: That raises a question which I wanted to ask. You picked out a primary objective or primary value of truth and you said, that there is secondary objective, which might be conducive to reaching the primary objective. But I guess I was curious. You did not say much about what you take the relationship to be between these cognitive values and non-cognitive values like Prof. Hick mentioned, "survival", even death, just perhaps being less miserable or being more miserable, or anyway, matters of that sort. So what do you think the connection to be in the case of belief between practical values and cognitive values?

KLEVAKINA-ULJANOV: I think the connection between practical values and cognitive ones is a kind of a means-end relationship. That is, if T is a goal and M is a necessary means for the realization of this goal, then a value which is attributed to M may be characterized as a practical value. Suppose, for instance, you put forward a simple scientific hypothesis which you believe to be an adequate one. Your goal is to achieve truth, whereas the means for this goal could be: acquisition of all scientific information available, a severe test, etc. Correspondingly, eternal and universal values like "survival" or "preserving health" could also be regarded as a means for some state of affairs. Would you not agree, then, that in a certain sense the means one chooses are secondary objectives which are conducive to the achievement of the primary aim?

LEHRER: I guess it's interesting. I think myself, that the interest in truth is of non-derivative character, i.e. it is an objective that people have quite parallel with practical objectives. But not everybody agrees with that. So some people think that the interest in truth is derivative from some other interests, for instance, in human welfare or something like that.

WEINGARTNER: When you spoke about reduction, you made the distinction between radical reduction, like an elimination of values and modest reduction, which is a kind of decreasing of values. I wanted to add that there is sometimes as we know from the history of science and of philosophy the case where new entities (which are positively evaluated) are introduced in order to solve some difficult problem. This is neither elimination, nor modest reduction. An example are Plato's ideas in order to solve the problem of universals. There are a lot of other such examples, say in mathematics, the introduction of sets by Cantor. In such cases, one could investigate, in what sense these new introduced entities have a kind of a value-impact on the problems considered. I mean, at least in the sense, that they are used as means for a solution to an important problem.

KLEVAKINA-ULJANOV: Yes, right, I am with you, but well, in any case I wanted to thank you for your comments, because my general purpose was just to draw attention to this problem, which seems to be not a trivial one.

HICK: Any further questions, comments, contributions?

SCHURZ: Maybe one remark. We spoke all the time about the question, whether values enter into reasons for believing and a very different question, which was also raised by you, that is, whether the content of a proposition or a proposition of a certain class always has a value-coponent: If this is so and very often it is so, a very interesting question for me is whether I can separate the value-component of a proposition from its descriptive part. I would say that as a methodological maxim, at least for a science, one should try to separate it. And what you said in the end of your talk presupposes the possibility of this separation, too. Just to clarify the point a bit more. As an example of a value-laden scientific statement you used the following claim: "Theory A is better unified than B". That's an important criterion for accepting theories which has a value-component, i.e. "better unified". But here it is obvious how to separate. You have the notion of "being more unified", which is descriptive, and you have the claim "to be more unified is good" which is prescriptive. And so you see what I mean. Very often you can separate the descriptive part of the proposition from the normative, evaluative one and the interesting question is whether you can do it always. Some philosophers have doubted that and made a point against it. So, would you admit the possibility of such separation?

KLEVAKINA-ULJANOV: In some cases, yes, but not always, I think. One has to strive for such a separation, or to put it in other words, one should try to become as precise as possible about the degree of valuation in one's judgement. Accordingly, with respect to a scientific inquiry, one should try to elicit all value judgements that are made or might have been made. But there is no garantee, in my view, that one would always succeed in this enterprise, that is, in separating a descriptive part of a proposition from a prescriptive one.

LEHRER: I have a question. You raised a problem, which fascinates me and it has to do with simplicity. So one thing you said suggested that you took simplicity to be a secondary goal, truth being a primary one, which suggests that you thought, simplicity or an interest in simplicity is something that serves the interest in truth. But there are two views about the interest in simplicity. One is, that the reason we seek simplicity is just that we like simplicity. The alternative account is that the reason we seek simplicity is that we think that it leads us to truth. I think the second view is almost certainly wrong. It has almost never been the case, that the simple theory turned out to be true. I mean, the Copernician theory is not true. If you look at his theory, everything is much more complicated than this theory suggests. And so if you look at the history of science, then it seems to me, that the claim that simplicity or an interest in simplicity leads to truth, is false. That does not mean that we should ignore simplicity. Maybe the best we can do is to look for simple answers. So I was curious about your reaction, or anybody else's reaction.

KLEVAKINA-ULJANOV: Well, I did not intend to mean that simplifying the matters would necessarily help to attain truth. Again, I was looking for value judgements in scientific inquiry and, with this point in mind, I considered how consistency and simplicity as evaluative criteria functioned in the choice between the heliocentric and geocentric systems. In a certain sense, the Copernician theory was simpler and this was the fact which contributed to the victory of Copernicanism, although, generally speaking, simplicity is not a cardinal point for attaining truth. At the same time, I would say, that with the notion of simplicity everything is not quite clear to me. Thus, in his striving for truth, a scientist *could consider* simplicity to be a necessary requirement. And such an account would raise up the question: how evaluative a scientific inquiry is if a scientist makes simplicity a secondary objective in his striving for truth as a primary goal?

HICK: Another way you talk about simplicity as in terms of Ockham's razor, namely that, if a theory introduces, so to speak, a fifth wheel that does not do any work, then you should eliminate it and you are thereby gaining greater simplicity. But of course on the other hand one should never assume that what one is trying to understand is simple.

WEINGARTNER: Yes, I agree to throw out redundancies is also a very important aspect, but another aspect is: simplicity yes, but not for the price of truth. So, truth is always, I think, much more important than simplicity. And I agree with you that the matter is not so simple. At least the history of Natural Sciences shows that in fact the closer we come to comprehensive truth, the more complicated the theories become. Just look at biology, at physics, at chemistry, everywhere.

HICK: But you would agree, perhaps, that where you have an unnecessary complication in your theory, you should eliminate it?

WEINGARTNER: Yes, unnecessary complications should be eliminated. But another aspect worth mentioning is that simplicity has different meanings. Sometimes one can mean just less parameters or a simple form of an equation, but a completely different view is something which proved always to be a criterion of a very high value: searching for simplicity in the sense of searching for a unifying idea. This is what Einstein stressed so very much in his biography. Think of his unification of mass and energy or space-time or of Maxwell's unification. The new famous equations of Maxwell for the electric and magnetic forces are more complicated than that of Faraday but are unifying so far unrelated areas in physics.

SCHURZ: A short remark to simplicity. I think, one aspect of simplicity is, how it relates to uniformity, that is, if you think about points of a numerical value and you look for the simplest curve or something like that. Or even take a probabilistic induction, that's a kind of simplicity, too. And I think, one can argue reasonably that assumption of uniformity in nature is a guide to truth.

PAUL WEINGARTNER

SIMILARITIES AND DIFFERENCES BETWEEN SCIENTIFIC AND RELIGIOUS BELIEF

0. INTRODUCTION

According to the common view scientific belief (i.e. belief in scientific hypotheses or theories) has nothing to do with religious belief (i.e. belief in the creed of some religion). Though this common view is very widespread, it is usually not reflected upon or analyzed in some detail. One of the points in my paper is that on a closer look and deeper analysis it will turn out that there are a number of interesting similarities between the two. And also the differences show interesting and important characteristics. But to show that there are similarities and that there are differences is only one important task. It will become clearer on a more detailed analysis that both "similarity" and "difference" should not be taken too literally. Since in many cases a particular similarity will show some new differences in special details and a particular difference will show some new similarities on a deeper level.

1. SIMILARITIES

1.1 In both scientific and religious belief three important components can be distinguished:

(1) What is believed, the content of belief.
(2) The believing itself or the action of believing.
(3) The reasons for believing.

Three questions correspond to these three components:

(1') What is believed?
(2') Which is the scientific or religious activity?
(3') Why is it believed?

Epistemic Logic deals mainly with (1) and (2). And only these two components enter the axioms and theorems of epistemic logic. Thus theorems like $B\neg p \rightarrow \neg Bp$ or $Bp \rightarrow \neg B\neg p$ concern (1) and (2) together, whereas distribution-laws like $B(p \rightarrow q) \rightarrow (Bp \rightarrow Bq)$ show the relations between (1) and (2).

P. Weingartner (ed.), Scientific and Religious Belief, 105–142.

Treatments in Philosophy of Science deal with (3), but sometimes underestimate the importance of (1) and (2) and their relations.

The three components play an important role in both scientific and religious belief. It will be shown subsequently that there are important questions coming up in both kinds of belief which are concerned with these components.

1.2 In both scientific and religious belief the content of belief is of propositional character.

That is, what is believed, is something, what is the case (or is not the case) something what is true (or is false).

Although this may seem to be trivial at first side it was and still is a problem at least in religious belief. There are many discussions of that question among leading theologians.[1] I agree with Bocheński that in real religious discourse at least a part – and in fact an essential part – of what is believed is propositional, i.e. is something which is (the case) or is not (the case), something which is true or false. When I say "in fact an essential part", I mean, that this part is concerned with the "creed" of that special religion; i.e. with those selected propositions, which characterize a particular religion and which have to be believed by someone in order to be counted as a member of that religious society or as belonging to that special religion.

Though there is no big problem concerning the propositional content of what is believed in scientific belief, there is a further similarity concerning the religious creed. Also in a special scientific discipline there is something like a creed, a selected set of propositions, the members of which have to be believed by someone who is counted as a member of say the physicists, biologists or mathematicians. Such a creed consists usually of the well confirmed hypothesis and theories and of the well established laws and theorems in that discipline plus a number of background assumptions which also transcend the discipline. One may also add certain interpretations of crucial experimental results. Great scientists sometimes do not believe everything belonging to the creed of a discipline (at a certain time) especially, if it is concerned with some basic assumptions. This leads usually to new theories (think of Newton, Maxwell, Einstein or Planck etc.) This shows also that there is an important difference between the religious and the scientific creed: the former does not change in its basic assumptions, the latter can change even in its basic assumptions (more on that in chapters 2.2 and 2.5).

1.21 In both scientific and religious belief the content of belief is also of normative character.
That is a part of what is believed is something what ought to be the case (or ought not to be the case), something which is valid (or invalid) (i.e. something which is also evaluated in a two-valued way). Although this claim may seem to be strange or wrong in the first place, it will become more evident with some examples.

It is essential for a religious believer, that he not only believes in certain states of affairs, but also in the validity of norms. Thus it is essential for a Christian believer to believe in the validity of the ten commandments or more general in the validity of any norm ordered (and revealed) by God. The same is true of the great other religions like Judaism, Islam or Buddhism.

But there is also a similarity in science here. It is also essential for a scientist to believe in the validity of certain methodological norms. Otherwise he cannot work. Thus general norms like "Base your hypothesis on all the information available" or "The new theory should include the correct results of the old theory (as special cases)" are accepted by all scientists whereas special ones like "Physical laws should be invariant against coordinate (space-time) transformations" will be believed at least by scientists of that special discipline.

Also in respect to norms there is a similar difference like that concerning the creed. Methodological norms in science are sometimes revised. Thus the norm to look for continuous dependencies of causes and effects (in order to be able to describe them by differential equations, the so-called lex continuitatis) was a successful methodological norm from Newton's time on, but was severely restricted by quantum theory.

Norms in religion may also be interpreted according to contemporary needs, but it seems, that they cannot be revised so radically as some of them have been revised in the development of science. Though to be just, one should compare of course very general and basic norms on both sides on the one hand and more specifically applicable norms on the other. And concerning the belief in the first the difference in science and religion might not be so great.

1.3 Concerning both scientific and religious belief, one has to distinguish (at least two) different concepts of belief.

These two different concepts of belief may be called the knowledge-exclusive belief (abbreviated as G-belief) and the knowledge-inclusive belief (abbreviated as B-belief). The former (G-belief) is characterized by the condition, that if someone believes something, then he does not know it and if he knows it, he does not (or need not) believe it. Whereas the second (B-belief) is characterized by the condition, that if someone knows something, he also believes it, but if he does not believe it, he also does not know it.[2]

1.31 Examples for G-belief: Before the proof of the independence of the Continuum Hypothesis (from the axioms of set theory) was given, v. Neumann believed (but didn't know), that the Continuum Hypothesis is independent. After Gödel proved the first part[3], i.e. that the General Continuum Hypothesis (GCH) can be consistently added to the axioms of Neumann-Bernays-Gödel-Set Theory (even if very strong axioms of infinity are used), v. Neumann wrote: "Two surmised theorems of set theory, or rather two principles, the so-called 'Principle of Choice' and the so-called 'Continuum Hypothesis' resisted for about 50 years all attempts of demonstration. Gödel proved, that neither of the two can be disproved with mathematical means. For one of them we know that it can not be proved either, for the other the same seems likely, although it does not seem likely, that a lesser man than Gödel will be able to prove this".[4]

But after the proof of the second part – that also the negation of GCH can be consistently added to the axioms of Set Theory (it holds for both systems, that of Zermelo-Fraenkel and that of Neumann-Bernays-Gödel) – was given by Paul Cohen in 1963[5] – von Neumann didn't any more believe it, but knew, that GCH was independent (from the axioms of Set Theory).

The General Theory of Relativity (completed by Einstein in 1915), made three important predictions: (a) The perihelion of Mercury, (b) the deviation of light rays which pass close to big masses and (c) the red-shift of the light reaching us from distant stars. The first (a) was known as an effect (not explainable with Newton's Theory) before Einstein's Theory was created. And so the prediction (and explanation) because of stronger gravitation since Mercury is much closer to the sun by General Relativity was a success immediately. In this case Einstein knew the positive result of the test of his theory. In the case of (b) and (c) he strongly believed, that they are correct and there will be a positive test possible too. In 1919 came the first confirmation of the prediction (b), the deviation of light rays: A British expedition of astronomers observed a total eclipse of the sun in Africa and confirmed the effect, that light rays from a star which run very close to the sun are deviated towards the sun (in general towards great masses). Later better and more exact confirmations of (b) were obtained.

In 1922 the Soviet meteorologist Alexander Friedmann believed (but did not know) and predicted (on the basis of Einstein's picture of dynamic space), that the entire universe is in dynamic change. In 1929 the American astronomer Edwin Hubble confirmed this prediction. He found out, that the light reaching us from distant stars, is shifted towards red of the spectrum (red-shift) and that this red-shift is proportional to the distance of the emitting star(s) (galaxy). This red-shift has a simple explanation as the Doppler-Effect caused by the recession of distant stars (galaxies) away from us. This was the confirmation of (c) which was later again confirmed many times. Thus after these positive results of testing the predictions Einstein knew, that predictions (b) and (c) are correct and are positively confirmed by tests. And this means also, that he did not and needed not believe (G-believe) that anymore, since there is sufficient justification to say that he knows now.

1.32 Examples for B-belief: No special examples for B-belief are necessary since B-belief may be interpreted in the following way: To B-believe that something (p) is the case means just to think that p is true (valid), to hold that p is true (valid), to strongly assume that p is true (valid) etc. Thus if someone knows that chromosomes duplicate, then he also B-believes it and also if someone G-believes that GCH is independent (from the axioms of Set Theory), then he also B-believes it.

1.33 Religious belief is always knowledge-exclusive, i.e. is always first of all G-belief. Since if one believes religiously – for instance that Christ came for the salvation of mankind or that there will be some kind of conscious life after death – one does not know it (and knows that one does not know it). And this holds for all religious beliefs even if not necessarily for all the statements of the creed of some special religion. Since first of all the statements of the creed might not be all logically independent of one another such that some believer may infer one proposition of the creed from some others. And in this case he knows that one is a consequence of the other. Such inferences may be also done by theological argumentation. Still the propositions so derived are not known but believed, as known consequences of others which are believed.

Independently of that some theologians who studied some field carefully might know something what other religious believers only believe. Some passages of Thomas Aquinas show that some of those problems have been already discussed (seven hundred years) earlier.

"Any science is possessed by virtue of principles known immediately and therefore seen. Whatever, then, is an object of a science is in some sense seen. Since it is impossible that the

same thing be seen and believed by the same person, as shown already, it is also impossible that the same thing be an object of science and of belief for the same person.
What may happen is that the same point seen or known in a science by one person is believed by another. For example, we have hope that we will one day see what we now believe about the Trinity; We see now through a glass darkly, but then face to face. (I Corinthians 13,12) That very vision the angels already enjoy; thus what we believe in they see. Similarly, even in this life it can happen that what is an object of vision or science for one person, can be an object of belief for another who does not have proven knowledge about it.
Matters set before the whole human community for belief, however, are in no instance the object of any science, and these are the object of faith pure and simple. In these terms faith and science are not about the same object." [6]

"Or else the arguments derive their cogency from principles of faith, i.e. scriptural authorities, as Dionysius points out. On the basis of such principles some point stands proved for believers in a way similar to that whereby for all men something stands proved on the basis of principles known connaturally." [7]

"As Aristotle says in the same text, (Post. An. 89b2) some people can have science, others opinion about the same point; this we have also established about science and faith. In the case of the one person, however, while he may have faith and science about a matter that is the same broadly speaking, i.e. it relates to the same subject, he cannot with regard to a point that is the same strictly speaking. In other words it can happen that someone has science about one facet of a subject matter but opinion about some other facet. Similarly, the one person can have apodictic proof that God is one, and belief that he is triune. But when it comes to the same thing under the same aspect, science cannot coexist in the one person with either opinion or faith, although for different reasons. Science cannot, because by definition science means that one judges that the object known cannot possibly be otherwise than it is; opinion by definition means that its object is thought of as able to be otherwise than it is. As to faith, because of its certitude, it also judges that what is believed excludes the opposite; but the reason for the incompatibility between the same thing's being an object of science and of belief is that what is known is seen and what is believed is not, as has been established." [8]

But what generally holds is that if somebody religiously believes something, then he does not know it, but he holds it to be true, he strongly assumes it to be true. It seems to me that this also holds for religious belief in important norms which are written down in religious texts like the Bible, the Koran or the Sutta Pitaka in Buddhism. Important and ultimate norms like those proposed of religions are not known. The religious believer believes that they hold (are valid) and gives a justification with the help of other beliefs: (1) God has revealed (required) it. (2) And when God reveals something, then it is true or it holds. Or: (1) God has given us (by revelation) certain rules for acting in society. (2) When God has given a rule, then it should be obeyed.

1.34 G-Belief, B-Belief and Knowledge

What the examples show are important relations between the three. They can be shown in the following scheme

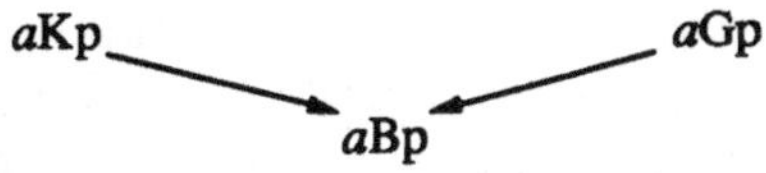

Here '*a*Kp', '*a*Gp' and '*a*Bp' stand for '*a* knows that p', '*a* G-believes that p' and '*a* B-believes that p'.

Further important propositions are the following ones:

1.341	$aKp \rightarrow aBp$	$\neg aBp \rightarrow \neg aKp$
1.342	$aGp \rightarrow aBp$	$\neg aBp \rightarrow \neg aGp$
1.343	$aGp \rightarrow \neg aKp$	$aKp \rightarrow \neg aGp$
1.344	$aGp \rightarrow \neg aK\neg p$	$aK\neg p \rightarrow \neg aGp$

1.4 Both scientific and religious belief have several different relations to knowledge.

1.41 First there is the important difference between knowledge and belief: Knowledge is sufficiently justified at least for the present purpose. Belief – i.e. both scientific and religious belief – is not sufficiently (but only partly) justified. There is always somewhere a gap, a lack in the justifying reasons. How is the gap filled?

The gap cannot be filled by sufficiently justified reasons, otherwise it would be knowledge and no more belief. But it can and is often filled by evaluations which again do provide only partial justification. A scientific hypothesis may be evaluated higher than another one because of different reasons; more objective criteria are that it has passed severe tests, explained more facts, and in more detail, has unified so far unrelated concepts ... etc.[9]

More subjective reasons (though not only subjective) are that the hypothesis was proposed by a famous or successful scientist (for instance by a Nobel Prize Winner).

Concerning religious belief there is the important meta- or background-belief that anything which was revealed (said, ordered ... etc.) by God is of higher value than anything else together with the belief that a certain proposition (belonging to the creed of a religion) was revealed (even through and with human mediators) by God. Other evaluative reasons are those which connect a certain belief with the last end in the sense of making it a necessary condition for the last end, for the eternal happiness. (More on that in chapter 2.4.)

1.42 Further relations to knowledge in both scientific and religious belief:
1.421 If the person *a* G-believes that p (is the case), then *a* knows that he believes. This is true for both scientific and religious belief provided the belief is conscious.
1.422 If the person *a* G-believes that p (is the case), then *a* knows what it is what he (*a*) believes. This is also true for scientific and religious belief provided "*a* knows what it is what he believes" is not taken too accurately. Believers may differ considerably in this respect, the one analyzing the consequences and impacts of his beliefs, the other just being able to give a more straightforward and simple answer when asked what it is what he believes.
1.423 Though 1.343 holds *a* may G-believe some proposition p at time t_1 and know p at a later time t_2. This can be true in scientific belief – recall the examples given in 1.31 – but may be true also in religious belief. Thus for instance St. Paul says in the Christian Bible that when we will know later (in the eternal life) our religious beliefs will have an end.[10]
1.424 Though again 1.343 holds person *b* may G-believe something of a certain thing and know something else of the same thing. Thus in particle physics it often happens that some of the properties of an elementary particle (say a new discovered meson) are known because of certain experiments, but other properties are only believed or conjectured. In the history of science there are of course a lot of such examples.

The famous example in the history of Christian philosophy and theology is here the existence and the essence of God. The former we can know, the latter we cannot know in this life according to great philosophers of the Middle Ages, from Clemens of Alexandria through Thomas Aquinas.[11] Concerning the second then we have to believe what revealed texts tell us partially and in parables and analogies.
1.425 Finally it is of course not incompatible with 1.343 that someone *a* G-believes that p and someone else *b* knows that p. This situation might arise in science when some scientists have already a proof or a confirming test or experiment for something of which others have still conjectures and belief.

In religion this is said when religious believers are compared to those who have reached the last end like the saints who already know what in this life can only be believed.

1.5 Both scientific and religious belief satisfy consistency criteria.

1.51 I do not suggest here that either scientific or religious belief has strong conditions of logical closure which would imply logical omniscience or deductive infallibility: Certainly I do not claim that someone who G- or B-believes something (scientifically or religiously) believes also all the logical consequences of what he believes. Nor that any believer would believe in all the logical valid propositions. These properties are not properties of human knowledge or belief, they may belong to higher intelligent beings. It is therefore also not an adequate explanation, if it is said, that these concepts are ideals since such properties as omniscience concerning logic and deductive infallibility are beyond human capacity. Why do logicians or mathematicians derive theorems and succeed in making difficult proofs, if these theorems are known by rational knowers?[12]

Such an understanding of epistemic concepts, like those of knowledge and belief comes from constructing epistemic logic by reinterpreting modal logics such that for instance the necessity operator is taken to be the operator "knows that". Though there are similarities, like $\Box p \rightarrow p$ with $Kp \rightarrow p$, there are striking differences: all logical consequences of necessary propositions are necessary, but not all logical consequences of known propositions (there are infinitely many) are known again. The result of Solovay has shown that '$\Box$' can be interpreted in an interesting way by the concept "provable" (in arithmetics) and this shows again clearly that an interpretation with "know" is not adequate because of the differences between "know" and "provable".[13]

For these reasons (and for others which cannot be discussed here) the following criteria of consistency are modest.

1.52 Consistency Conditions

Since I understand belief as conscious belief (not as a disposition to believe), I claim for both scientific and religious G-belief, that it is impossible to believe (at the same time) that p and that non-p. In other words, if person *a* believes that p, he does not believe that non-p and if he believes that non-p, he does not believe that p. This also holds for B-belief:

1.521

$aGp \rightarrow \neg aG\neg p$

$aG\neg p \rightarrow \neg aGp$

1.522

$aBp \rightarrow \neg aB\neg p$

$aB\neg p \rightarrow \neg aBp$

From this it follows, that it cannot be the case that *a* believes that p and *a* believes that $\neg p$ (at the same time). Also *a* believes that both p and $\neg p$ is impossible and usually even stronger than the above if one assumes a distribution law only in one direction: $aG(p \wedge q) \rightarrow aGp \wedge aGq$.

In fact these laws are very general and hold also for the operators "K" (knows that) and stronger ones like "provable that" or others like "obligatory that" in Deontic Logic or "wills that" in the Logic of (rational) willing.

It is important to realize that the opposite arrows do not hold. For instance $\neg aGp \rightarrow aG\neg p$ would claim that if *a* does not believe that p, he believes that non-p. This might be sometimes the case (i.e. for some propositions p), but certainly not always, otherwise the principle $aGp \vee aG\neg p$ (*a* believes that p or *a* believes that non-p for any proposition p) would hold in general, which is certainly not the case. This analogous disjunction does also not hold for human knowledge, i.e. it does not hold that (for any p): $aKp \vee aK\neg p$ though it is a consequence of omniscience (i.e. of: $(\forall p)(aKp \leftrightarrow p)$). For the concept "provable" it does not hold in general either, i.e. it does not hold that for any p: $Prov(p) \vee Prov(\neg p)$; but if it holds in respect to a certain system of propositions (closed sentences) we call the system (syntactically) complete. For "obligatory that" or for "wills that" it does not hold either. The latter, i.e. "for any p: wills that p or wills that non-p" is – if applied to God – the thesis of the all-willing God, which has no foundation in any great religion, but is attributed to God by Calvinism, religious fatalism and is implicitely presupposed by those who attribute to God the willing (and causing) of every evil.[14]

The fallacious argument in question goes usually as follows: (1) If God wills that evil does not occur, then (because he is almighty) evil does not occur. (2) But in fact evil occurs. (3) Therefore God wills that evil occurs. Though (1) and (2) are true (3) does not follow from (1) and (2) because by modus tollens it follows correctly only (4): it is not the case that God wills that evil does not occur. And it is a logical fallacy to derive (3) from (4) since (4) is not a double negation; just like it would be a fallacy to derive (5): the physician has proved that he has cancer from (6): it is not the case that the physician has proved that he has not cancer. Likewise it would be a fallacy to derive (7): p is provable from (8): it is not provable that non-p. Also (9): A is good does not follow from (10): it is not the case that A is not good (bad). Because if some things are neutral they are neither good nor bad and therefore (9) does not follow from (10). And since some propositions are neither provable nor disprovable (7) does not follow from (8). And if the physician has not yet proved (or not yet finshed the proof) about someone's cancer (5) does not follow from (6). And so also for some evil neither God wills it nor he does will that it is not the case, i.e. he keeps back his will from it or does not interfere (and this he wills). And that is all what is derivable from (1) and (2), but (3) is not derivable except one presupposes as a further premise the thesis of the all-willing God which is not a doctrine or claim of any great religion.

There is, however, the further question whether in science or religion we can have it that someone believes that p at one time t_1 and believes that $\neg p$ at a later time t_2. I think one can accept such cases. That is I do not claim consistency of beliefs (neither scientific nor religious) of a person through time. But I do not introduce time indices here, because they are only needed on some special places and otherwise would only complicate the matter (cf. 1.423).

There are other possible Consistency-Criteria, but the main question is, whether they are not too strong. It seems to me that the following two principles can be accepted for both scientific and religious G-belief.

1.523 $\quad \mathrm{Con}(a\mathrm{G}p \wedge p) \rightarrow \mathrm{Con}(a\mathrm{G}p)$
1.524 $\quad \mathrm{Con}(p) \rightarrow \mathrm{Con}(a\mathrm{G}p)$

(where 'Con(p)' stands for 'p is consistent')

The reason for accepting 1.523 is the general idea that if a conjunction is consistent, then also its parts must be consistent. The reason for accepting 1.524 is this: If the content (p) of a belief (*a*Gp) is consistent, then p can have roughly three evaluations: logically true, factually true, factually false (one may even add positive probability-values). In the first two cases that is if person *a* believes in something which is true, it is certainly correct to say that his belief is consistent. But since the third case is a fact – in the sciences we often have beliefs in something which is (turns out to be) false – it must be consistent to believe in false propositions. Therefore 1.524 holds too.

Certainly one cannot accept the principle

1.525* $\quad \mathrm{Con}(a\mathrm{G}p) \rightarrow \mathrm{Con}(p)$.

Frege's belief in his axiom of comprehension – $(\exists y)(\forall x)\,[x \in y \leftrightarrow Fx]$ – refutes it. He says in his 'Vorwort' that a controversion could arise only about this principle (axiom V) though he believes in it as a principle of logic.

"Ein Streit kann hierbei, soviel ich sehe, nur um mein Grundgesetz der Werthverläufe (V) entbrennen, das von den Logikern vielleicht noch nicht eigens ausgesprochen ist, obwohl man danach denkt, z.B. wenn man von Begriffsumfängen redet. Ich halte es für rein logisch."[15]

In his 'Nachwort' he says:

> "Einem wissenschaftlichen Schriftsteller kann kaum etwas Unerwünschteres begegnen, als dass ihm nach Vollendung einer Arbeit eine der Grundlagen seines Baues erschüttert wird. In diese Lage wurde ich durch einen Brief des Herrn Bertrand Russell versetzt, als der Druck dieses Bandes sich seinem Ende näherte. Es handelt sich um mein Grundgesetz (V). Ich habe mir nie verhehlt, dass es nicht so einleuchtend ist, wie die andern, und wie es eigentlich von einem logischen Gesetze verlangt werden muss."[16]

Though Russell showed (by making an instance with '$\neg(x\exists x)$' for 'Fx') the axiom to be inconsistent, it should not follow that Frege believed (consciously) an inconsistency, what certainly wouldn't be true. By the principle 1.525*, however, we would be forced to say that Frege believed (consciously) in an inconsistency, which is false though it is true that he believed something which was in fact inconsistent but without being aware of it. The reason for such situations is that the properties of human rationality are such that not all (logical) consequences of what one believes one either knows or believes too. And this is because one neither knows nor believes all the propositions $q_1, q_2, \ldots q_n$ which are derivable from p, i.e. $p \vdash q_1, p \vdash q_2 \ldots p \vdash q_n$ for all q_i derivable from p if one believes in p. This means that there are cases in which although the content of a belief is inconsistent, the belief itself might be consistent. And this is so for two reasons: First in the sense that every belief is a fact and facts are real and therefore possible. Second in the sense that the inconsistency in the content is not transparent (hidden) for the believer such that the content is for him not incompatible to any logical principle or even to any other belief he has.

As an illustration of an even stronger case we might consider the so-called indirect proof: One wants to prove $\vdash q$ and for the sake of the proof assumes $\neg q$. In this case we have even to accept that aGq and $a\text{A}\neg$q ('A' for 'assumes that' or 'entertains that'; the latter was proposed once to me by R. Chisholm) at the same time. This of course presupposes a rather weak operator A for which the analogous laws of 1.521 and 1.522 do not hold. Although from aGq it certainly follows that aAq.

1.6 Both Scientific and Religious Belief satisfy certain simple important logical principles.

I do not give here an exhaustive list of such principles which certainly would be also rather difficult. Since the amount of principles available to a believer in his background-knowledge certainly will depend on his education, field of interest and area of research. But there are some very simple ones which are used – besides the usage in everyday life – quite frequently in scientific and religious discourse.[17] Of these latter ones I'll give a non-complete list.

1.61 Modus Ponens
$[a\text{Gp} \wedge a\text{G}(\text{p} \rightarrow \text{q})] \rightarrow a\text{Gq}$

1.62 Modus Tollens
$[a\text{G}\neg\text{q} \wedge a\text{G}(\text{p} \rightarrow \text{q})] \rightarrow a\text{G}\neg\text{p}$

1.63 Hypothetical Syllogism
$[a\text{G}(\text{p} \rightarrow \text{q}) \wedge a\text{G}(\text{q} \rightarrow \text{r})] \rightarrow a\text{G}(\text{p} \rightarrow \text{r})$

1.64 Disjunctive Syllogism
$[a\text{G}(\text{p} \vee \text{q}) \wedge a\text{G}\neg\text{q}] \rightarrow a\text{Gp}$

1.65 Dictum de Ommi
$a\text{G}(\forall x)\text{F}x \rightarrow a\text{G}(\text{F}a)$

1.66 $[a\text{G}(a = b) \wedge a\text{G}(\text{F}a)] \rightarrow a\text{G}(\text{F}b)$

1.67 The principles 1.521

1.68 $a\text{G}(\text{p} \wedge \text{q}) \rightarrow (a\text{Gp} \wedge a\text{Gq})$

Concerning principle 1.68 one could defend – for the serious scientific and religious believer who reflects on what he believes – also the version with the opposite arrow. This point is dealt with in the discussion.

It will be plain that from 1.61 with the help of the exportation-law we also have

1.69 $a\text{G}(\text{p} \rightarrow \text{q}) \rightarrow (a\text{Gp} \rightarrow a\text{Gq})$

I assume that all these laws hold also for the weaker kind of belief, for B-belief.

1.7 Scientific and Religious Belief accepts some meta-principles about belief and believers.

1.71 All believers believe something which is true. Or: Nobody believes only (exclusively) false things.

$$(\forall x)(\exists p)(xGp \wedge p) \qquad \neg(\exists x)(\forall p)(xGp \rightarrow \neg p)$$

Certainly every scientist will grant this to any other scientist. But I think also mature religious believers will grant this to other religious believers even if they do not belong to the same religion. The view that all the claims of the other religion are false, seems not to hold under serious religious believers who know something of the other religion at all. Moreover it is not difficult to show that there are some common beliefs in the creed of – at least all main-religions (Judaism, Christianity, Islam, Buddhism, Hinduism). One example would be that this life here is not the ultimate end of man such that the Christian principle "What shall it profit a man if he shall gain the whole world, and lose his own soul"[18] seems to be accepted in an analogous way in all main-religions. One such common belief is sufficient however to make principle 1.71 true.

1.72 For no believer his creed is complete.
By the creed (cf. 1.2) of a believer I mean a system of propositions (maybe in addition norms or rules) which characterize a particular religion (or scientific discipline, here the respective basic laws and very well confirmed principles and hypotheses) which are believed by someone who is counted as a member of that religious society or as a researcher in a special scientific discipline. I call such a creed complete in respect to a special field (with which the creed is concerned) if all true propositions (and all valid norms or rules) of that field are contained in the creed or are logically derivable from it. It is not difficult to see that 1.72 holds: A person whose creed would be complete is an analogon to a omniscient being (in respect to a certain field) where knowledge is replaced by belief. Since this is a rather extreme conception, the following consideration seems to be more rewarding: Is the creed sufficiently complete for a certain (important) purpose? Applied to the religious believer: Is the creed (of some special religion) sufficiently complete to reach the eternal life – according to that religion? This question I think can be answered in most cases (in respect to most religions) with yes. In respect to scientific belief analogous questions are: Is the creed of a scientific believer (in respect to a discipline or field of research, for example astronomy) sufficiently complete to support a certain hypothesis or theory (say about the age of the universe)?

1.73 All believers believe at least one thing (something) that is false. Nobody believes only true propositions.

$$(\forall x)(\exists p)(xGp \wedge \neg p) \qquad\qquad \neg(\exists x)(\forall p)(xGp \rightarrow p)$$

Today's scientists will not have difficulties to accept 1.73, though there might have been difficulties in the time of Galileo to accept fallibility for all scientists. From the history of science we know that 1.73 is true and therefore also very plausible for the future.

Concerning religious belief one has to distinguish the doctrine of a religious institution as for example the Catholic Church and the individual members belonging to a certain religion. In general the religious institution would not so easily accept 1.73. Concerning the Catholic Church there is the case that Pope John Paul II admitted officially before the scientific world that the Catholic Church had a false belief in the controverse with Galileo. Even if this debate – after we were taught the general dependency on the point of reference from Relativity-Theory – has lost (or should lose) its excitement. As to the individual believers it depends whether they are critical and mature. In this case I think they would accept 1.73 knowing that by reflecting on their religious beliefs they have certainly revised some of them, developed their understanding for others and found out they were mistaken in respect to some or misinterpreting them. 1.73 does not say which beliefs are false. Thus it is compatible with 1.73 that no religious believer would admit that his basic beliefs (the ones concerning the gist of the creed of some religion) could be false.

1.8 Both scientific and religious belief have important reasons for belief.

As was pointed out already in 1.4 the reasons for believing are never completely sufficient, otherwise we would speak of knowledge. However, there seem to be important reasons which are common to both kinds of belief and which are not sufficient, but necessary conditions.

1.81 That what is believed is not impossible.
For both scientific and religious belief it holds that the believer B-believes also that that what he believes is not impossible, i.e. consistent. And in both cases it seems to be a task of the respective discipline to find out whether this is the case. Since if the believer not only B-believes but also knows that what he believes is not impossible, this is an important support (reason) for his belief.

1.811 $a\mathrm{Gp} \rightarrow a\mathrm{BCon(p)}$

1.812 $a\mathrm{Gp} \rightarrow a\mathrm{B}\Diamond\mathrm{p}$

That that what is believed (scientifically or religiously) is impossible or inconsistent is a serious hinderance to believe it. This is not only so in science but also in religion and theology; the latter at least according to great theologians such as Anselm of Canterbury, Albert the Great and Thomas Aquinas. Therefore it is a task of theology – according to them – to show that what is proposed for faith (the gist of the beliefs of a religion which are comprehended in the creed) is not impossible or inconsistent: "Arguments advanced by the saints to prove matters of faith are not strict proofs but plausible reasons, in that they show that some point of belief does not involve a contradiction".[19]

That the consistency of what is believed is an important component for scientific investigation has been pointed out by scientists and philosophers ever since. A particular example is Leibniz who required to show the consistency for a notion introduced into science beyond Descartes' criterion clare et distincte: "La Regle Generale que plusieurs posent comme un principe des sciences, quicquid clare distincteque percipio est verum, est sans doute fort defectueuse."[20]

> "Je viens à vostre Examen du grand principe des Cartesiens et de Dom Robert, que j'ay déja touché: scavoir que nos idées ou conceptions sons tojours vraies. Et comme j'ay déja dit, je suis bien eloigné de l'admettre, parce que nous joignons souvent des notions incompatibles, en sorte que le composé enferme contradiction. J'ay examiné plus distinctement ce principe dans une remarque sur les idées vrayes ou fausses que j'ay mise dans le journal de Leipzig. Et je tiens que pour estre asseuré, que ce que je conclus de quelque definition est veritable, il faut scavoire que cette notion est possible. Car si ell implique contradiction, on en peut conclure en même temps des choses opposées."[21]

It is known that in the mathematics of the beginning 20th century Hilbert proposed his program of consistency proofs for mathematics. This program had two steps: The first was to formalize a part (field, area) of mathematics and the second to establish a metamathematical theory with finite means in which the consistency proof of the formalized part is to be carried through by finite methods. Hilbert believed in this program in such a way as to hope to solve every mathematical problem with it. It was shown that this program cannot be carried through not only because consistency proofs require much more than Hilbert thought, but what is more important for our problem that even a proof of consistency does not (also not in mathematics) guarantee truth.

Moreover 1.811 is different from 1.812 since '$\Diamond$' cannot generally be interpreted as 'Con' (consistency). Since though for a formalized theory T, $\mathrm{T} \rightarrow \Diamond\mathrm{T}$

is certainly true and provable within T (if T includes ◇ and axioms at least as strong as the weak modal system of Feys), T → Con(T) is in general not provable in T (according to Gödel's results).

Also in the empirical sciences consistency plays an important role. New hypothesis should be consistent with well confirmed laws and other well established hypotheses. The set of laws of a theory have to be compatible among themselves. Inconsistency is a hinderance for belief in general. A new hypothesis which is incompatible with well established laws will hardly be believed by scientists. But there are cases in which a new hypothesis was incompatible with certain presuppositions in the foundations (for instance concerning space and time), but it included well established laws as special cases (like Einstein's Special Relativity).

1.82 That what is believed has consequences which are confirmed.

The claim of 1.82 is almost trivial in most cases of scientific belief in hypotheses and laws. Hypotheses often start (genetically) from generalizations of data which are then interpreted with the help of the hypotheses and the data are at the same time confirming instances. Usually a lot of other confirming instances (where not only the number, but also the quality and severeness of tests is important) are necessary to strongly believe in the new hypothesis. However, there are exceptional cases where a new hypothesis cannot be tested at the time it is proposed, though it predicts tests of a certain kind. Then considerations of type 1.81 will be made first. Later it may then be possible to test it: Thus Newton's general gravitation among masses was tested only by Cavendish (1798, 111 years after the first edition of Principia). Similar with General Relativity-Theory (cf. 1.31).

But does 1.82 also hold for religious belief? I think it does. Religious believers have a view in which they interpret the beauty and order in the world as a confirming instance for their belief in an almighty and omniscient creator. Or they see confirming instances of ethical doctrines in the bible when observing family life or life in society. By these ethical doctrine I do not mean pure norms but certain predictions about life which are connected with norms. For instance like in the fourth commendmend. There are then also 'negative' confirmations in the following sense: If the ten commendments or the principle of charity are frequently violated in a particular society and if there is also disorder, hate and a lot of corruption in that society, then these facts are interpreted as confirmations in the validity of the violated principles. Even if this latter idea belongs more to the Christian and Jewish Religion the following idea is more general: The general experience that a too high evaluation of and concentration on outer passing things of this world leads men to restlessness, sorrow and unhappiness is interpreted as confirming the general doctrines of all religions to higher evaluate the supernatural inner and lasting goods of mankind.

2. DIFFERENCES

2.1 In religious belief that what is believed is already there, laid down in a text like the Bible or the Koran or by oral and ritual tradition and is often summarized in the 'Creed' of a religion. In scientific belief that what is believed is proposed, conceived, invented, constructed within the development of that scientific discipline.

In the introduction I said that also within the differences one can find similarities. Thus in the discussion it was pointed out by John Hick that at the beginning of a religion there is no such established corpus like the 'Creed' nor a text or an oral and ritual tradition. This is correct and thus a comparison of what is believed at the beginning of a religion with what is believed at the beginning of a developing scientific discipline may show again more similarities.

But if we compare established religion and established scientific discipline, then the above mentioned difference is more intelligible. Though, granted, there is also a kind of 'scientific creed' consisting of presupposed basic assumptions – and this again is a similarity – there is a considerable part in every science which changes and developes: New hypotheses are permanently developed which describe and explain old or newly discovered phenomena. New hypotheses supersede and correct older ones and sometimes completely new ideas come up and change a lot in the interpretation of known facts.

2.2 Religious belief includes the belief that what is religiously believed cannot be false. Scientific belief admits that what is scientifically believed can be false.

This claim needs some clarifying comments.
2.21 First concerning religious belief. What is meant here is certainly serious belief in important propositions of a creed of a religion. The creed of a religion consists of those propositions by which a believer belonging to that specific religion is defined, i.e. as someone who believes those propositions of the creed.[22]

But concerning such propositions it holds that if they are believed religiously, this belief includes (implies) the belief (the strong assumption) that those propositions cannot be wrong. Symbolically:

$$2.211 \qquad aG_Rp \rightarrow aB\neg\Diamond(aG_Rp \wedge \neg p)$$

For some religions, especially Christianity, Judaism and Islam the reason for this lies partially also in the fact that the religious believer believes that those propositions (of the creed) have been revealed by an omniscient and almighty god even if this revelation happened via mediators (cf. 2.3 and 2.4). The more mature religious believer would certainly concess that there might be theological (exegetical) discussion about the correct interpretation of some of the propositions of the creed (of a particular religion). But this does not mean that 2.211 is violated. One way of handling this is to understand 2.211 conditionally in the sense that the mature religious believer grants to make the presupposition "if p is correctly interpreted". But independently of that there are some basic beliefs which do not allow much debate of interpretation: either Christ (more than a prophet) was born or not, either Christ resurrected or not. Though there may be a debate on the meaning of Christ (son of God etc.), but hardly about "was born" or "resurrected". Similarly with "there will be reward or punishment after death" etc. And in respect to such propositions the religious believer also believes that they cannot possibly be (altogether) false, even if he may grant that he might not understand some detail here.

2.22 Concerning scientific belief I do not think that the analogue of 2.211 never holds. This was a point which was brought up in the discussion. The example I used in my talk was the law of the conservation of energy which says that in a closed system (and especially also in the universe as a whole) the amount of energy is constant. Concerning this law most physicists would say that they not only believe in it but also believe that it is impossible that this is false. Another basic belief in physics is the constancy of the light velocity with its wellknown amount (of ~ 300.000 km/sec) – at least relative to our universe.[23]

There are less good examples in mathematics. Because before a proof is given, the analogue of 2.211 would not hold for one of some famous conjectures in mathematics, say Goldbach's or Fermat's. On the other hand after a proof was given, say that the axiom of choice or the Generalized Continuum Hypothesis are independent of the axioms of Set Theory, one does not speak of belief anymore but of knowledge. In cases where a proof can hardly be expected we also cannot apply the analogue of 2.211. Thus only few would claim that the belief in the consistency of Zermelo-Fraenkel or Neumann-Gödel-Bernays Set Theory cannot be false.

The important difference between scientific and religious belief in this respect seems to me to be this: In a great many cases – which admittedly do not concern the most basic beliefs of a certain scientific discipline – scientific belief in a certain scientific hypothesis, law or theory admits that what is believed can be false. This means that one could have the opposite belief in the consequent of 2.211 if scientific hypotheses are concerned:

$$2.221 \qquad aG_s p \rightarrow aB\Diamond(aG_s p \wedge \neg p)$$

In the discussion[24] the question was raised whether 2.221 can be applied to conjectures in mathematics. Since we admit distribution-principle 1.68 also for B-belief and accept the distribution law $\Diamond(p \wedge q) \rightarrow (\Diamond p \wedge \Diamond q)$ it follows from 2.221 that $aG_s p \rightarrow aB\Diamond\neg p$. And the latter is problematic whenever p is proved (provable) in mathematics is interpreted in such a way that $\Box p$ holds (where '$\Diamond\neg$' is understood as '$\neg\Box$'). Thus let p be the conjecture of v. Neumann that the Generalized Continuum Hypothesis is independent of the Axioms of Set Theory. According to the passage cited (cf. 1.31) he believed that p is provable, i.e. that p can be proved or – according to the interpretation – that p be necessarily true. And the latter is incompatible with the belief that it is possibly false (i.e. with: $B\Diamond\neg p$).

To this difficulty I want to say three things. First of all we might interpret "*a* believes that p is provable" as "*a* believes that it is possible that p is proved" and so if proved is interpreted with the box we get "*a* believes that it is possible that p is necessary". But this is not incompatible anymore with the belief that p is possibly false (i.e. not necessary), provided that not strong Modal Systems are presupposed.

Secondly Standard Possible World Semantics (Kripke-style) has the necessitation only for truths of logic, i.e. if $\vdash p$ (where '$\vdash$' means provable in Logic, usually in First Order Predicate Logic) then $\Box p$. It is one of the strange consequences of this semantics that even very simple basic truths without which mathematics is impossible like $(\exists x)(\exists y)(x \neq y)$ (there are at least two individuals) are not provable with $\Diamond$ in this semantics, i.e. $\vdash \Diamond(\exists x)(\exists y)\, x \neq y$ is not a theorem. Needless to say that also $\Diamond(2+3=5)$ is not provable in this semantics. Thus if '$\Diamond$' in 2.221 is interpreted with the help of the usual Possible-World Semantics, 2.221 is not applicable to propositions (beliefs, conjectures) of mathematics.

Thirdly we might interpret ◇ and □ intuitively and accept the rule that what is proved (and provable) in mathematics should hold with □. Then 2.221 is not correct for mathematical conjectures but only for conjectures in empirical sciences in the sense that we view the propositions in these disciplines as contingent, at least contingent relative to those of logic and mathematics. Though also in these disciplines one might distinguish between propositions which are necessary like laws ('empirical necessity', 'natural necessity') and others which are not, this kind of necessity is certainly not the same as that of logical and mathematical truths. In order to have an analogous principle of 2.221 for conjectures in mathematics (and one may add: in logics) we might interpret '◇' in another way, for example epistemically as Keith Lehrer proposed in the discussion. Then 2.221 might read: If *a* G-believes (in mathematics, logic) that p, then he B-believes that it is compatible with all what he believes and knows that p might be false.

It seems, however, that in mathematics the following principles 2.222 and 2.223 are more important. As for the first we might consider again the conjecture of von Neumann (cf. 1.31) and similar cases in which it is believed that what the conjecture says will be proved later (i.e. is provable; here 'provable' is not interpreted as 'necessary'):

2.222 $(\exists p)\ [aG_S p \wedge aG_S(\text{Provable}\ (p))]$

But if the mathematician believes that p is provable he certainly will not believe that p is possibly false. Thus 2.222 might be an analogon for 2.211 in mathematics.

On the other hand there seems to be also an analogon to 2.221. As said before 2.221 holds for many cases of scientific belief outside logic and mathematics. But the following analogon of it seems to hold for situations where it is known (or strongly believed) that a proof is very difficult. Then mathematicians might believe that the proposition is true but unprovable:

2.223 $(\exists p)\ [aG_S p \wedge aG_S\neg(\text{Provable}\ (p))]$

There are cases in respect to which we know the result because a proof has been given. In this case one can speak of scientific knowledge and no more of scientific belief. For instance it is known that the consistency of a theory T cannot be proved (is not provable) within (with the means of) T.

To give a non-exhaustive list of the cases in scientific belief one might distinguish the following possibilities:

(1) Like in religious belief 2.211 holds if a very basic law in natural science is concerned.
(2) On the contrary, 2.221 holds if new not very well tested hypotheses are concerned or are even a theory which has some special difficulties.
(3) There are many cases in between like beliefs that a hypothesis is a better explanation of the phenomena than another, but nevertheless might be false or that it is a very good approximation to truth but might still in the future lead to some difficulties (false consequences).
(4) Contrary to (1) sometimes even very basic principles or basic presuppositions are admitted or believed to be possibly false. Example: The assumption about space and time in Newton's Theory and in Einstein's Theory.
(5) For some situations in mathematics 2.222 as an analogon to 2.211 or 2.223 as an analogon to 2.221 may hold.

2.3 Religious belief is connected with the belief and the desire in a state of happiness or reward after this life (in this world or – for most of the religions in another one) whereas for scientific belief there is no such connection with a transcendent state. Therefore religious belief is connected with desire and will in a special way which is not the case for scientific belief.

The particular connection of religious belief to the transcendent state of happiness or reward which is conceived as the final goal is this: If a person believes religiously that p – and – provided that p belongs to the Creed of that religion, i.e. to the important statements of it – then he also believes that if he would not believe that p, he would not reach the state of happiness or reward (his final goal) (cf. 2.31) And since the religious believer believes in the existence of such a state of happiness (in the actual existence concerning saints or prophets and in the possible existence for him) and further since he wants to be happy (cf. 2.32), he wants to believe. To arrive at this consequence we certainly need a principle concerning human will which was presupposed. It is this: If *a* wills that p is (will be) the case and *a* believes (G- or B-) that q is a necessary condition of p, then *a* also wills that q is (will be) the case (cf. 2.33). If 'q is a necessary condition of p' is interpreted as 'p → q' (of Standard Classical Propositional Logic), then similar paradoxes as the Ross-Paradox in Deontic Logic and others may arise. In order to avoid them one can add relevance criteria as I have proposed elsewhere.[25]

The principles which are involved in order to derive the conclusion are these:

2.31 $aG_Rp \rightarrow aB_R[(\neg aG_Rp \rightarrow \neg H(a))]$
'$H(a)$' for a is (will be) happy (rewarded)

2.32 $aWI[H(a)]$
i.e. for arbitrary religious believers a, a wills (wants) to be happy (rewarded) (in a lasting sense) after this life

2.33 $[aWIp \wedge aB(p \overset{r}{\rightarrow} q)] \rightarrow aWIq$
'$(p \overset{r}{\rightarrow} q)$' for '$p \rightarrow q$ according to some of the proposed relevance criteria'

From 2.31 – 2.33 one can deduce 2.34 which states one important feature of the will in religious belief:

2.34 $aG_Rp \rightarrow aWI[aG_Rp]$

Principle 2.34 states an important point. Many great christian philosophers, at least from Augustine on stressed the important impact of the will in religious belief. For them religious belief (faith) was understood from the very beginning as an action of both intellect and will. This action belonged to both sides of the highest sphere of the human soul, e.g. to reason which included both intellect and will.

Concerning principle 2.34 I want to state an analogy between religious and scientific belief. Although in scientific belief there is no analogous principle to 2.31 still 2.34 has some analogon. It does not hold generally as in religious belief but it may hold sometimes for particular cases that if some scientist believes in a scientific hypothesis, then he also wants to believe in it. Think of the inventor of some new interesting hypothesis. Of him it may be true to say that he also wants and wills to believe in the hypothesis he has proposed.

2.4 Religious belief searches for or tries to provide or claims ultimate and global explanations. Scientific belief does not claim and usually neither searches nor tries to provide ultimate and global explanations.

Ultimate and global explanations are such which provide a highest reason such that there is no need or there even is no possibility to give another higher reason for that. Thus an ultimate cause which causes everything except those actions for which men have free will or an almighty omniscient creator who created the universe are examples.

Though science too aims at powerful explanations and though hypotheses, laws and theories are replaced by more general hypotheses, laws and theories in the course of development of a certain scientific discipline and even by comprehensive theories interrelating several areas of disciplines or several disciplines, still ultimate explanations for all known phenomena (of any science pace of any experience) are not claimed nor offered nor aimed at. However, this holds of the scientific attitude of the 20th century and fairly well also of that of the newer time since Galileo. At the beginning – think of science in Antiquity especially when theoretical questions were discussed like in Greece – there were global questions (What- and Why-questions) and global and ultimate answers of the form: everything is such and such or everything was caused by such and such. And some philosophers continue to ask questions and give explanations in that style while scientists found out that at least for themselves this attitude is not rewarding.

On the other hand it was always understood as the task of philosophy and religion to give more general answers and more general explanations than the sciences will ever give. Thus the difference between science and religion is apparent if today's science is meant. But still I do not want to finish this chapter without mentioning some newer branches of science in respect to which this difference is again smaller. There is modern cosmology which tries to explain the beginning and development of the whole universe, there is the extension of Darwin's theory to the development also of non-living things and of the universe, there is general Systems-Theory which tries to interpret everything as certain ordered structures ... etc. Thus in these branches there are certainly trends which come closer to the characteristicum of religious belief to search for and to provide (if not to claim) rather global and very general (even if not ultimate) explanations.

2.5 Religious belief involves believing (in) a highest (omniscient and almighty) and credible authority whereas for scientific belief this is not essential.

Concerning 2.5 there are certainly serious differences between the so-called Abraham-religions: Judaism, Christianity and Islam on the one hand and Buddhism, Hinduism, Shintoism and other religions from the Far-East on the other.

For the latter group of religions 2.5 cannot be claimed in an unrestricted way, but for the three Abraham-religions it seems to be an important common feature.

Believing in a highest (omniscient and almighty) authority is connected with point 2.4. This authority is understood as the creator or cause of the world (universe) and so can serve as an explanation for all the phenomena in this world with the exception of those in respect to which men has free will. That there is such an area of free will in which men is consciously causing certain actions for which he is responsible is a belief of all the three first mentioned religions. Even if some extreme versions which deviate from the mainstream like Calvinism denies such an area.

Believing in a credible authority means believing in someone who tells nothing but the truth and never misleads somebody. The belief in such an authority has as supporting premises the belief that the authority can (is able to) tell nothing but the truth because he is omniscient and that he does so because he is good.

Concerning scientific belief there is no such single epistemic authority for all scientific believers. Though two things have to be mentioned here. First of all believing in epistemic authorities[26] is essential and necessary both for the learning process (if the young scientist studies his subject) and for the communication of scientific results. It is just not possible to repeat every scientific result, investigation or proof (or only those which are relevant to one's own research) for oneself. Secondly some epistemic authorities may have proved to be reliable (in respect to a restricted field) in a very high degree, they might have become famous scientists in a well justified sense. Then it is certainly more rational to believe them than to not believe them and also more rational to believe them more than those others whose reliability was discredited because of some or other incorrect claims. Thirdly there is so to speak the authority of the community, say of mathematicians, of physicists, of biologists, of historians etc. They (or most of them) believe and defend the well corroborated results, laws and theories of a special discipline, the results which belong to the kernel of a discipline, one could even in a sense speak of the "creed" of a discipline.

Thus there is also authority in this sense in respect to scientific belief and this shows again a similarity within the difference. For – in respect to the difference – one still has to remember that neither the authority in the learning process or in communicating scientific results nor that of famous scientists or that of the "creed" of a discipline is such that it would be believed to be infallible. And in fact in some cases some claims of even famous scientists turned out (later) to be wrong and also some essential parts of the "creed" turned out to be mistaken. As examples think of Newton's conception of absolute time and space (space as a container) which was replaced by Einstein's conception of curved space-time or of the universal applicability of the lex continuitatis (only continuous dependencies between cause and effect which can be described by differential equations represent the moving behaviour of bodies in the universe), which was overthrown for the area of microphysics by Quantum Mechanics. In this respect I agree essentially with Karl Popper's fallibilism for all sciences which stresses the provisional and temporary character of scientific hypothesis – even if an approximation to truth is possible – on the one hand and the importance for modesty and Socratic attitude of the scientists on the other.[27]

Thus although there is also belief in authority in the scientific belief there is no such belief in an absolute authority who tells nothing but the truth because he is omniscient and good.

2.6 For religious belief in Judaism, Christianity and Islam it is essential that it manifests itself in religious activity as a consequence of belief whereas for scientific belief there is no such strong connection.

More accurately, for the three mentioned religions it is not enough to believe. Acting as a consequence of belief and according to the belief is necessary, according to the passages: "Not those who say Lord, Lord, but those who obey the will of my father ..." "What you do to the least of my brothers ..."

Thus "living" or "active life" according to the belief in both religious statements and norms (for example ten commandments) is a necessary condition for religious belief at least according to the three Abraham-religions.

Though concerning scientific belief there is no such necessary condition, one must not forget, that "activity" is also required here: the proposed new hypothesis has to be exposed to severe tests, it has to be confronted with reality, it has to be used to explain already known and new phenomena ... etc.

Concerning religious belief one might say that the belief is not serious if it is not accompanied by active life according to the religious norms. And this – the belief and the activity – has to occur in the same person. But in science it may split up in the sense of division of labour. The theoretician might invent and propose the hypothesis and draw some conclusions from it, but the experimentalist might work on the procedure to test it.

In religion no such "division of labour" is possible. The religious believer cannot leave the religious activity, the religious life to the other.

2.7 At least Christian religious belief includes the belief that for believing religiously one needs the help of God in the sense of receiving God's grace.

This is especially the doctrine of the New Testament that without God's help and especially Christ's incarnation the believer cannot do anything, not even believe religiously in the right way, let alone being justified by living a just life.

Although we cannot say that for scientific belief an analogous help is needed, one still speaks of great scientists as blessed or endowed with great gifts or as highly gifted.

NOTES

1 Cf. the discussion in J.M. Bocheński (1965), ch.13, p. 40ff.
2 For G-belief cf. what Aquinas says about religious belief in his (STh) II-II, qu. 1 and 2.
3 K. Gödel (1940)
4 J. v. Neumann (1951), my italics, in: J.J. Bulloff - Th.C. Holyoke - S.W. Hahn (1969). The "Tribute to Dr. Gödel" from which the passage is cited was given by v. Neumann in March 1951 on the occasion of the presentation of the Albert Einstein Award to Gödel. It appeared in print in the volume *Foundations of Mathematics* (ed. Bulloff et al.), a collection of papers given at a symposium commemorating the sixtieth birthday of Kurt Gödel.
5 Cohen P. (1963/64). Cf. Cohen P. (1966).
6 Thomas Aquinas (STh) II-II, 1,5.
7 Ibid. ad 2.
8 Ibid. ad 4.
9 Popper gives such criteria of evaluation in his *Conjectures and Refutations*, p. 232 f.
10 Cf. the citations of Aquinas in chapter 1.33.

11 Cf. Clemens of Alexandria, (Strom) V; 12 and (Prot) 6. Augustinus, (Lib Arb) II, 15, 39; (Ord) 2, 18, 47; (Trin) I, 1, 1-2. Boethius (Con) III, 10; (Trin) 5,5. Thomas Aquinas (STh) I, 2, 3; 12, 11 and 13,1. Cf. the quote referred to in note 7.

12 In Weingartner (1982) I have criticized these kinds of epistemic systems and have proposed both a set of desiderata for epistemic concepts belonging to ***human*** rationality and a system of epistemic logic for the concepts knowledge, G- and B-belief and assumption.

13 Cf. Solovay (1976).

14 Cf. my essay on religious fatalism and the problem of evil (1974).

15 Cf. Frege (1893) Vol. I, p. VII.

16 Cf. Frege (1903) Vol. II, p. 253.

17 I do not mention trivial ones like the fact that scientific and religious believers believe in the law of identity x = x, in the law of non-contradiction (a proposition and its negation cannot both be true; this is a tolerant form which does not exclude many-valued logic) or in the multiplication table.

18 New Testament, Markus 8,36.

19 Thomas Aquinas (STh) II-II, 1, 5 ad 2.

20 Leibniz GP I, p. 384.

21 Ibid.

22 Cf. Bocheński (1965), §§ 3.3, 3.4, 13.1.

23 According to an interesting theory of A. Meessen the amount of the light velocity is dependent on the amount of energy distributed in the universe satisfying the equation $c = 2.a.E_u/h$ (where c is the light velocity, a the smallest measurable distance yet unknown, E_u the energy of the universe and h Planck's constant). If the amount of energy would be greater, the light velocity would be greater too, but still constant in that other universe. Cf. A. Meessen (1989).

24 Cf. the discussion of my paper in this volume.

25 Cf. Weingartner-Schurz (1986), Schurz-Weingartner (1987) and Weingartner (1988) and (1993). All the criteria proposed in these papers avoid the mentioned kinds of paradoxes. On the other hand the usual systems of relevance logic (cf. Anderson-Belnap (1975)) which provide a stronger → are much too weak to avoid any of those paradoxes.

26 For the concept of epistemic authority cf. Bocheński (1974), (1988). Bocheński defines authority as a threeplace relation: A is an authority of B in respect to the field C. If the field is a set of propositions he speaks of epistemic authority, if it is a set of norms of deontic authority.

27 Cf. K.R. Popper (1934, 1989), (1959), (1963), (1974), (1984).

REFERENCES

Anderson, A. R. and Belnap, N. D.: 1975, *Entailment. The Logic of Relevance and Necessity. Volume I*, Princeton University Press, Princeton and London.

Aristotle: (Post. An.) 'Posterior Analytics', in *Prior and Posterior Analytics*, intr. comm. W.D. Ross, Oxford 1957.

Augustinus: (Lib Arb) II, *De Libero Arbitrio, ML Vol. 32*, 1231-1310;
(Ord) 2,18,47; *De Ordine, ML Vol. 32*, 959-976; (Trin) I,1,1-2; *De Trinitate, ML Vol. 42*, 819-1098.

Bocheński, J. M.: 1965, *The Logic of Religion*, New York University Press, New York.

Bocheński, J. M.: 1974, 'An Analysis of Authority', in: *Authority*, F.J. Adelmann, S.J. (ed.), Boston College Studies in Philosophy, Bd. 3, Chestnut Hill/The Hague: Martinus Nijhoff, 56-85.

Bocheński, J. M.: 1988, *Autorität, Freiheit, Glaube*, Philosophia Verlag, München-Wien.
Boethius: (Con) III,10; *De Consolatione Philosophiae, ML Vol. 63*, 579-862;
(Trin) 5,5; *Quomodo Trinitas unus Deus ac non tres Dii, ML vol. 64*, 1247-1256.
Clemens von Alexandrien: (Strom) V,12; *Stromata*, O. Stählin (ed.), Vol. 1-6 (1906), Vol. 7-8 (1909); (Prot) 6; *Protreptikos, Paidagogos*, O. Stählin (ed.), (1905).
Cohen, P.: 1963/64, 'The Independence of the Continuum Hypothesis', in *Proceedings of the National Academy of Sciences* **50** (1963), 1143-1148 and **51** (1964), 105-110.
Cohen, P.: 1966, *Set Theory and the Continuum Hypothesis*, Benjamin, New York and Amsterdam.
Gödel, K.: 1940, *The Consistency of the Axiom of Choice and of the Generalized Continuum Hypothesis with the Axioms of Set Theory, Annals of Mathematics Studies 3*, Princeton University Press, Princeton.
Meessen, A.: 1989, 'Is it logically Possible to Generalize Physics through Space-Time Quantization?' in *Philosophy of Natural Science, Proceedings of the 13th International Wittgenstein Symposium*, Paul Weingartner and Gerhard Schurz (eds.), Hölder-Pichler-Tempsky, Wien, 19-47.
ML, Migne, Patrologia, series latina.
Neumann, J. v.: 1951, 'Tribute to Dr. Gödel' in *Foundations of Mathematics*, J. J. Bulloff, Th.C. Holyoke, S.W. Hahn (eds.), Springer, Berlin 1969.
Popper, K. R.: 1934, 1969, *Logik der Forschung*, Julius Springer, Wien/J.C.B. Mohr, Tübingen.
Popper, K. R.: 1959, *The Logic of Scientific Discovery*, (English translation of Popper (1934)).
Popper, K. R.: 1963, *Conjectures and Refutations*, Routlege-Kegan Paul, London.
Popper, K. R.: 1974, *The Philosophy of Karl Popper, Part Three: The Philosopher Replies*, P.A. Schilpp (ed.), Open Court, La Salle.
Popper, K. R.: 1984, *Auf der Suche nach einer besseren Welt*, Piper, München.
Schurz, G. and Weingartner P.: 1987, 'Verisimilitude Defined by Relevant Consequence-Elements. A new Reconstruction of Popper's Original Idea', in *What is Closer-to-the-Truth?*, Th. Kuipers (ed.), Rodopi, Amsterdam, 47-77.
Solovay, R.: 1976, 'Provability Interpretations of Modal Logic', in *Israel Journal of Mathematics* **25**, 287-304.
Thomas Aquinas (STh) I,2,3; 12,11 and 13,1. *Summa Theologica*, transl. by Fathers of the English Dominican Province, Christian Classics, Westminster, Maryland 1981.
Weingartner, P. and Schurz, G.: 1986, 'Paradoxes Solved by simple Relevance Criteria', *Logique et Analyse* **113**, 3-40.
Weingartner, P.: 1988, 'Remarks on the Consequence-Class of Theories', in *The Role of Experience in Science*, E. Scheibe (ed.). De Gruyter, Berlin, 161-180.
Weingartner, P.: 1993, 'Can there be Reasons for Putting Limitations on Classical Logic', in *Mathematical Philosopher*, P. Humphreys (ed.), Patrick Suppes, Kluwer, Dordrecht and Boston.

DISCUSSION

KUTSCHERA: You have introduced two notions of belief, which was G-belief and B-belief. We have a principle, that says, if a G-believes that p, then a does not know that p. Now for B-beliefs you have generally, if a knows that p, then he believes that p. If you don't B-believe that p, then you don't know that p. So a Gp means that a B-believes that p and does not know that p.

WEINGARTNER: Yes, aGp implies aBp and $\neg a$Kp.

KUTSCHERA: For otherwise, if you do not distinguish these two kinds of belief, you would have an easy proof for scepticism: if we know, we believe, but if we believe, then we do not know, so we don't know.

WEINGARTNER: Yes, I agree. In this sceptical argument 'belief' is used in two different meanings (B-belief and G-belief) and so the absurd consequence is derivable.

HICK: What I am going to say is very minor, suggesting some qualifications in some of the differences that you listed. (1) You say that in religion what is believed, that is the propositions believed, are already there, whereas in science they are constructed as science goes along. This would be true of an established version of religion, no doubt, with its formed creed and doctrines. But what about the early days of a tradition, before a system of doctrines has been formed? Then it would not be quite so true perhaps. (2) And then, second, you say that in science one believes that what one believes may be false, but in religion not. I would have thought, that if you have extremely basic beliefs, in each case they are not held with the thought that they might be wrong. For example, an extremely basic belief of a scientist is that the world is there to be explored and understood. This is not, I think, revisable. And perhaps corresponding with that, something like "There is a transcendent dimension to the universe" is not revisable in religion. On the other hand, specific religious doctrines, like God is three persons in one, are discussable and there are Christian people you know who give it a very minimal meaning, such as that God has many aspects. So I am inclined to think that a theological doctrine is in effect a theory and is revisable, just as is a scientific theory. On the other hand, of course, if you say that doctrines are revealed and have divine authority, then the situation is different. But notice that the idea of revelation, and of a form of words having divine authority, is peculiar to certain religious traditions and does not occur in others. For example, in Buddhism it is totally a foreign idea, and in Hinduism it's a foreign idea in that the Vedas are not said to be revealed by a personal God. And then again the idea that it is characteristic of religious beliefs that they must be held, and if one does not hold the right ones, one will fail of salvation, is also specific to certain forms of Christianity. So what I am pointing out is that the analysis you have given is based on Christianity, and not only on Christianity but perhaps specifically on Catholic Christianity, and that if one took a broader view of religion some of these differences might disappear and perhaps other ones might come forward.

WEINGARTNER: Thank you very much for the interesting additions. I want to comment on some of them. First, it is quite right that you say, there are basic beliefs in science of which most of the scientists think, that they are really established and one example I have given in my paper was the law of conservation of energy. This is thought to be so well confirmed and established, that if new hypotheses come up, they have to be compatible with it. I agree with you in the sense that as far as the most basic beliefs are concerned there is an important similarity rather than difference between scientific and religious beliefs. I want to add something what I often criticize in my Philosophy of Science lectures: it is the view of the homogeneity of a scientific discipline; one picks out some important sentences and then analyses them and investigates its properties and then thinks to know what the beliefs of this discipline are. This is a great misunderstanding. Thus $2 + 2 = 4$ is completely different from let's say the axiom of choice. In the first case we can admit strong evidence. It is basic in the sense, that in no mathematical theory such statements as those of finite number theory could come out false. For the axiom of choice however we cannot assume such an evidence and also not for other axioms of set theory or for instance for the continuum hypothesis. Similar in physics: The law of conservation of energy or the constancy of the light velocity (at least relative to the energy amount in this universe) are quite good examples of very basic beliefs, whereas the string theory in quantum mechanics is an example where scientists would admit it could be possibly false.

My second comment is concerned with your point that theological doctrines are revisable. I'd like to distinguish beliefs of religious believers and theological doctrines or theological theories. And I agree with you that the theological doctrines and theological theories are revisable and in fact have been revised through history. One interesting case is the history of dogmas in the Catholic Church. They originate often so, that some theological theories have been competing for some time (sometimes for centuries) and finally some was ruled out, by showing, that it clashes with some statements of the Bible or with some exegetically well established points of the creed. There was a similar view for doctrines of Christian Philosophy. The compatibility with statements of the Bible as a revision-principle for Christian philosophy was already proposed by Clemens of Alexandria.

But with the beliefs (of religious believers), which concern parts of the creed, it is not so, I think. Although there might be some discussion concerning interpretation (exegetically), the propositions of the creed seem not to be so revisable.

Thirdly you are quite right to say that before a system of doctrines concerning some religion is established I should not say that what is religiously believed is already there. Thus when the early Christian Church discussed which scriptures are canonical – they have done this by first laying down some rules and methods – one cannot so easily speak of an established creed, though some more primitive creed which was passed over from one generation to the other was probably there too.

BOCHENSKI: I shall take the liberty of asking Prof. Hick a question about Buddhism. I always understood, that the very term "Buddhism" comes from the name of Buddha, who is the eluminator, a man, who teaches. But if I understood you, you said, that in Buddhism there is no revealing personality.

HICK: The Buddha regarded himself as having discovered by experiment in meditation – an interesting analogy with science, – an experienced reality and also certain truths. But other people are to believe them, not because he has taught them but by rediscovering them for themselves.

BOCHENSKI: As a matter of fact there is a sort of trust to Buddha or there is none? Do they trust him like a Moslem, who trusts Mohammed?

HICK: I don't think so, no. He himself did not want people to believe anything because they trusted him to tell the truth. Rather they should follow the path, the practical path, and discover the truth for themselves.

LEHRER: I have a couple of technical objections. First to your claim that in scientific belief, if one believes that p, then one also assumes that p could be possibly false. I don't think, that correctly describes the state of von Neumann when he believed in the sense of G-belief, that the Continuum Hypothesis was independent of Set Theory. He believed that and he did not know. But I don't think, that he believed, that it was *possible,* that he believed that and it was false. The reason being that this proposition is such, that, if it is true, it is necessarily true. Because if the proposition is such that it is either necessarily true or necessarily false, like the one that von Neumann accepted, he does not believe that it is possibly false. So I think your proposal does not capture propositions that are necessarily true or necessarily false, but it is a plausible account for contingent propositions. And I suspect in fact, that the modal operator, the possibility operator there really is not a logical modality, but an epistemic modality. Then if you read the right hand formula as saying *a* believes that for all he knows, he believes p and it is false, then of course it is okay. So I guess, if the modality is interpreted logically, it is not plausible, if it is treated epistemically, it is plausible.

WEINGARTNER: This is a good point. You want to say that my claim that scientists admit that their beliefs could be possibly false, is applicable only to contingent propositions, say in physics, but would not be applicable to mathematics.

Concerning v. Neumann there is also the passage of his "Eine Axiomatisierung der Mengenlehre" of 1925: "Denn wir wissen ja bloß, daß die bereits erledigten Sätze der Mengenlehre aus ihm folgen. Die unerledigten aber, z.B. das Kontinuum-Problem, könnten (wenn die Kategorizität fehlt) in einem ihnen genügenden Systeme richtig, im anderen falsch sein. D.h., es wäre überhaupt unsicher, ob diese Axiome zur Erledigung z.B. des Kontinuum-Problems hinreichen." (v. Neumann: Collected Works Vol. I, p. 54). Let me give another example: Gödel conjectured (strongly believed) in 1933, that the class of closed prenex quantificational formulas with prefixes $\forall\forall\exists...\exists$ and arbitrary predicate letters and identity sign, but no function signs or individual constants is finitely controllable (i.e. every satisfiable formula in it has a finite model). Dreben asked Gödel in 1966 for substantiation of the conjecture and Gödel replied, that he does not recall details, but remembers that the extension (from the class without identity, for which this result was proved by Gödel in 1933) provides "no difficulty". By 1970 it was generally agreed by people working in this field, that the extension provides serious difficulty and that it is an open problem (they believed Gödel's conjecture could possibly be false. And in 1984 Goldfarb proved that it was false (cf. Journal of Symbolic Logic 49, 1984, p. 1237 ff.).

LEHRER: I am sure there are such kinds of cases where scientists think seriously that the conjectured proposition could be false. I mean, we have got a good understanding here, but I think, this sense of '*could be wrong*' really is not captured very well by the *possibility-operator*, because with respect to mathematical propositions, if it's possible, that they are false, then they are false.

WEINGARTNER: Let me go back to my last example: What does it mean that "by 1970 it was generally agreed ... that it is an open (unsolved) problem". Your suggestion "for all he knows, he believes p is false", can certainly be used here: For all what experts in the field knew, the negation of Gödel's conjecture was (also) compatible with that knowledge. And concerning the possibility operator you are right, that it cannot be the usual one defined in the usual Kripke-style-Possible World Semantics: Because there it holds that if something is provable with $\Diamond$ it is also provable with $\Box$.

But I doubt that the concept "either necessarily true or necessarily false", i.e. "possibility implies necessity" can be easily applied outside logic, say in mathematics. The Continuum Hypothesis (in our example) or the axiom of choice has been proved to be consistent with Set Theory. Does something follow for its necessity or even for its truth therefore?

KLEVAKINA-ULJANOV: At the very beginning of your talk, you stressed, that you used the term belief, whereas Prof. Lehrer used the term acceptance. I mean, as truth oriented. And then you differentiated this weak belief B and strong belief G. And then my question is: How do you then differentiate this G-belief and B-belief with regards to scientific belief and religious belief? I mean, you still said before, that it is necessary to differentiate. When you speak about scientific belief, you would like to differentiate G-belief and B-belief. And when you speak about religious belief, you also differentiate this strong and the weak concept. What is the criterion?

WEINGARTNER: I pointed out, that there is an important similarity, that both, the strong scientific and the strong religious belief imply not knowing and secondly that they both imply a certain kind of weaker belief, which says something, like 'I think, that it is true or I accept it as true, I hold it to be true.' But what it means further, has to be discussed I think by lying down certain axioms or theorems which hold for both scientific and religious belief. Thus I think, that even for the weak belief (and the more for the strong) we have consistency criteria, e.g. that, if *a* believes p, then he does not believe non-p. And further for both kinds of belief modus ponens would hold in the given form and the distribution principle given in my paper. But some of those principles – for instance the consistency condition – do not hold, and this I want to add now, for a very weak kind of assumption as it occurs in an indirect proof. You assume p and want to prove it indirectly and therefore you also assume non-p, just for the sake of the proof. It might be even correct to say, that one G-believes p, but assumes for the sake of the proof – that non-p. The operator A (assumption) then does not obey the consistency criteria, i.e. it does not always hold, that if I assume p, then I do not assume non-p. Whether I need such a weaker kind of assumption also for religious belief, I am not sure. Thus in this case we might not have an analogy.

LEHRER: Let me ask you a question about the modus ponens principle there: $[aGp \wedge aG(p \rightarrow q)] \rightarrow aGq$. Again that does not seem to me to be plausible for this reason: it can be the case that, though a person believes p and believes that p implies q, he does not put them together, i.e. does not somehow see the connection between the two. The more plausible principle it seems to me would be: $aG(p \wedge (p \rightarrow q)) \Rightarrow aGq$. I think that's the one that is plausible, the one of your talk. The one you've got seems to me not to be necessarily true. Suppose, that I am a person who wants to avoid believing q, that somehow it is painful for me to believe q. When I think about p, I don't think about the fact, that p implies q. I do in fact believe p, and, though I do in fact believe that p implies q, I never put the two together, so I never draw the conclusion that q.

WEINGARTNER: This, first of all depends on the distribution rules. If there is an equivalence, then you can replace one modus ponens principle by the other. In my talk I had only the arrow instead: $G(p \wedge q) \Rightarrow (Gp \wedge Gq)$. Then the principle you suggest has a stronger antecedens. Secondly I agree that this situation may happen sometimes and also therefore one should not have too strong distribution principles. But I think, that the scientific believer and the religious believer are believers, who reflect on their beliefs. I have presupposed, that both are serious believers and are thinking over their beliefs and know by reflection to some extent at least what it is that they believe. Thus in science, if somebody says, I do not draw the conclusion, I do not put the premises together, this seems strange to me. The modus ponens is such a frequently used rule, that it would be a little bit strange if somebody really believes in p and also believes, that p implies q, without believing q.

LEHRER: Okay, that would strike me as a good argument for having a biconditional in that distribution principle. Because now it sounds like you think, that the scientific person, who believes p and believes q, also believes (p and q), I mean, it looks like you are requiring that the scientific person puts together his premises.

WEINGARTNER: Usually I think he does.

BOCHENSKI: I am not pleased with the English term 'belief' which you used. The meaning is so broad and that somewhat blurs the differences. On the one side, to say, I believe, that 2 + 2 = 4, is quite correct in English. But you won't say in German: "Ich glaube, daß 2 + 2 = 4", you say, "Ich weiß es". There is an opposition between 'glauben' and knowledge, you see. So my first remark would be this: What you say about belief is very correct as long as you talk about my religious hypothesis. But faith is something quite different. I should say, that my religious hypothesis is of the same level as the scientific hypothesis, with only the differences, which I tried to elaborate. But faith is something completely different. It has an absolute certainty and there are various other differences. One of them I thought to be general, but now I learned from Prof. Hick, that I am wrong as far as Eastern religions are concerned. I think, it is still true of the three mediterranean religions, Mosaism, Christianity and Islam: The scientific belief is a diadic relation between a person and a proposition. *A* believes *p*. But faith, in the meaning of those religions, is a ternary relation. There is the *believer*, the *proposition* believed and the revealing *prophet*. There is the relation of trust to the founder of religion, to the prophet. Prof. Hick says that it is not true for Buddhism. But there is also a difference inside Christianity. While Protestants tend sometimes to forget the relation to the proposition believed, the Catholics tended to neglect that linking the believer with the revealer. To Aquinas, however, belief (faith) is quite clearly at least a ternary relation. And it results from my conversations with religious believers, that the main point of their faith is the relation of trust to the revealing author, in our religion to Christ. I think, something very similar will occur in Islam, the trust to the prophet Mohammed.

WEINGARTNER: I have of course not excluded this revealing factor, because I said, that these propositions I believe religiously, are elements of the creed or are consequences of the creed and the believer knows or again believes this.

BOCHENSKI: Of course you are right concerning the religious hypothesis. The basic dogma says, everything which is included in the creed of the given religion, is true. But the psychological situation of the believer is, that once he is inside, he does not accept the creed because of the religious hypothesis, but because of the trust in Christ or in Mohammed.

WEINGARTNER: I agree, that the trust in the religious authority is a very important factor. Could it not come under the reasons for the religious belief? Reasons for believing (scientifically or religiously) are of course a very important component of both scientific and religious belief (faith), as I stressed in chapter 2.4 of my talk. But I still do not see the necessity of having a ternary relation.

BOCHENSKI: I don't think this is a reason for believing. The reasons for believing, I guess, would be the religious hypothesis, because it helps you organizing your experience or so. But once you are inside, then the structure is that described by Aquinas, namely a ternary relation. And I was even astonished to meet quite distinguished scholars, who asserted as a primary element in their religion the relation to Christ as a revealing factor.

LEHRER: Again it seems to me that's characteristic of certain religions, but not of all. Thinking of one's own happiness is not the only reason that one might believe. For instance, some might find certain experiences or certain arguments really compelling and come to belief as a result of them. Now, *that might not make them truly religious men in the sense that Hick has tal*ked about. But suppose it does. Now suppose, in addition, that these persons believe that they have sinned badly and won't be saved. Well, religion does not offer them much hope of happiness. On the contrary, if they have awfully sinned, they would believe that they will be doomed. So, though I am sure it is in fact a motive for a great number of religious believers, I am not sure myself that it is an essential ingredient in belief or even in faith.

One other question was the divine inspiration. But of course there are some scientists, who are believers, who also think, that their scientific inspiration is a form of grace. They then pray in gratitude for those ideas having come to them. This suggests to me that they think they are scientific believers inspired that way. They might be wrong, but I am not sure if that is inconsistent with being a scientific believer.

WEINGARTNER: I agree, that the connection with happiness is not the only reason for believing. I have discussed also others like consistency since inconsistency would be a hinderance. And all these reasons can never be understood as sufficient "premises" to determine someone to believe. But they are still very important and they can be necessary conditions for religious belief. And so I still think that it is at least a necessary condition for religious belief, this connection with happiness, or more generally with some other life later, where there will be reward, or punishment. I do not know, if this is so in all the religions, but John Hick might say something to that.

HICK: I would think. On the one hand there are – particularly perhaps in Calvinist history – cases of people for whom religion has been a matter of despair, because they have believed that they are damned and doomed. There is still a connection, as you say, with happiness, but it's a negative connection. The idea of a happiness they are going to miss is an essential idea, but it operates negatively instead of positively.

BOCHENSKI: If I understand you correctly, the collective happiness, not necessarily that for the individual. The transit from the Hinayana to Mahayana comprehends the assumption of a different attitude towards personal salvation. While personal salvation is essential in Hinayana, in Mahayana it is no more. You forget about personal happiness. It is perhaps happiness of the collectivity, of humanity, not your own.

HICK: Right, yes.

BOCHENSKI: There is a Dominican convent in Unterlinden, which I quoted today. When the sister superior was asked, what they would do, if there would be no immortality, the sister answered: 'Exactly what we do now.' You see, I think, there is a transition from primitive religion, which is completely oriented to personal interest in some way to the great religion as it's quite clear in Mahayana. In the Bible try to see the transit from the Exodus from the primitive religion of fear and personal interest up to the prophets and then to the gospel, then you will see there are many, many different forms of religion, and how they are not necessarily connected with personal happiness.

HICK: It is extremely complex, isn't it, because on the one hand, in very early forms of religion the notion of the individual had hardly been developed at all, and one thought of oneself simply as part of the collective, as so to speak a cell in the tribal organism. But certainly the great trust of Buddhism and aspects of Hinduism is to transcend the ego. So there is no thought of the ego, the conscious self, as acquiring an eternal happiness, but rather you are to transcend the separate self; and yet that transcending is itself in another sense a happiness. You find happiness in a different way.

WEINGARTNER: May I make a comment also to the second point of Keith Lehrer. I wrote in the written draft of my paper that scientists sometimes also believe that if they have invented a new interesting hypothesis, then this is also some kind of grace. My point would be only, that this happens from time to time and cannot be generalized and thus is not a necessary condition for scientific belief. Whereas there is the question, whether it is a necessary condition for religious belief. I think it is a necessary condition of Christianity, because of the parables of the vine(stock) and the tendrils and others, but I do not know about other religions. Is there such a thing that one needs also help from God in order to believe religiously?

LEHRER: I think, it would hold also for Judaism.

HICK: I would not think of it in connection with Judaism, nor with most forms of Buddhism, but of course Islam and Buddhism occur in the plural, and there is one form of Buddhism, Amida-Buddhism, in which you call on the name of Amida and are thereby open to divine grace. So it's a very complex picture of differences and similarities.

JOSEPH M. BOCHENSKI

RELIGIOUS HYPOTHESIS REVISITED

In a book on "The Logic of Religion" published in 1965 I formulated my views on what I called "the religious hypothesis" (RH). What I mean by that utterance is, by the way, different from what James means by it, as it does not denote in my text the same as "religious faith". In the present paper I shall briefly restate what was said in my book concerning that theory and then formulate and discuss doubts I have and new insights I believe to have had since that time.

I.

Here is a re-statement of my views as of 1965:

I assumed as certain that many men and women, among them some individuals highly respected for their rationality and intelligence, did and do accept religious beliefs by faith. One may like or dislike that fact, but it cannot be denied. Alongside with that, I assumed that while the acceptance of a faith requires a 'will to believe', to use James' expression, i.e. an irrational factor, I asserted that no mentally sane person can accept as true a proposition which she deems to be important to her, without some reason. But as religious beliefs are surely such propositions, it was assumed that there is some rational justification of the acceptance of a religious belief. In Catholic theology such justifications are known as *preambula fidei*.

Thus the question arose: what is that justification? That problem has often been treated the wrong way. Some authors tried to avoid it by assuming that religion has no propositional content, that it is just an emotional attitude; others refuted the rational factor in faith without a previous analysis of what it is. The first is obviously false on factual grounds; the second is contrary to the elementary principles of scientific method. Analysis is needed.

In the book I discussed three theories concerning the said justification: the deductivistic, the authority-theory and the theory of RH. I discarded the first, because it makes religious faith become a science, which it surely is not. I said that both of the others may be used by the believers – I thought, however, that only children and members of primitive compact societies can use authority for the justification of their faith – not the intellectuals. For them the only possible justification is that by the RH.

The RH orders the whole of the believer's experience – not only factual propositions, but also value-propositions. There are, consequently, two differences between it and a scientific hypothesis. First, its utter generality; second,

P. Weingartner (ed.), Scientific and Religious Belief, 143–160.

the fact that it is not intersubjectively verifiable, among other reasons because it also covers the said value propositions. It is also distinct from faith, as it lacks certainty, which is proper to every authentic religious faith.

Another important point in my older views is, that a man converting to a religion does not first accept single dogmas, but the meta-dogma of the given religion, namely the proposition saying that the whole content of the respective creed is true.

II.

So much about the views explained in my earlier book. During the twenty-five years that followed its publication several doubts and new ideas occured to me concerning the matter. I shall mention here four among them, concerning, respectively, the concept of religion, religious experience, authority and prediction.

The first, and perhaps the most important point concerns the simplification I imposed on the concept of religion. I supposed that the word "religion" denotes *religious systems* and nothing else. I think, by the way, that many authors writing on religion commit the same mistake.

This is how it can be shown that I was wrong. According to my theory, a man does not convert to the *dogmas* of the given faith, but to the its *meta-dogma*, which asserts that the given creed is true. He will, consequently, accept first, that (say) the Muslim creed is true and only later (at least logically later) that God exists – because the latter is a dogma of his faith and not its meta-dogma.

Now experience contradicts this. It is namely an empirical fact that many converts first accept the existence of God, i.e. a dogma and not a meta-dogma of their creed.

This can perhaps be answered by saying that the term "religion" may refer to two different, if interconnected, objects: on one side to what I call "basic religion", on the other to the religious systems:

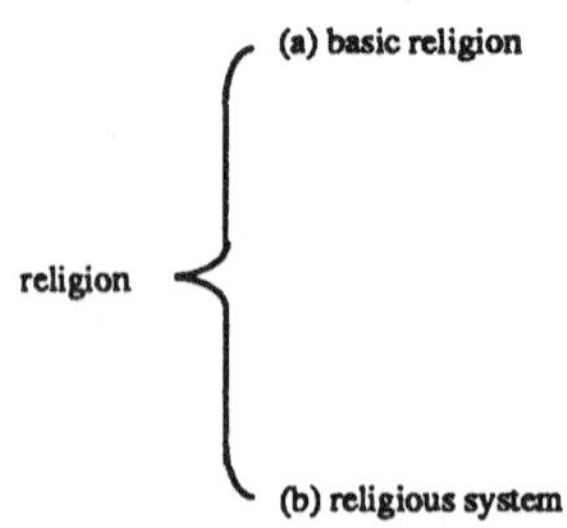

By "basic religion" I understand the fundamental religious *Weltanschauung*. I am using the word "*Weltanschauung*" here in the primitive Diltheyan meaning of the term. Actually it has two different meanings: the Diltheyan and the modern. According to the latter, a *Weltanschauung* is a complex, sometimes even sophisticated system like those of Communism and of National Socialism. But in the Diltheyan meaning the *Weltanschauung* is an emotional attitude toward reality, the propositional content of which is rather small. Thus the proposition or propositions, making out the creed of the basic religion could be e.g. "the world is guided by Providence" or "there is something sacred". By "religious system" I mean creeds such as those of the great book-religions.

The conversion of a person who was previously non-religious would then come in two steps. The first consists in the acceptance of the basic religion and the second in the acceptance of the meta-dogma of a specific religion system, as described in my book.

And the conclusion I draw in regard to the theory of RH is that both types of religion should be treated separately. It may be that the logical structure of the acceptance is different in each case. As far as I can see the situation is now the following.

The theory of RH does still apply to the religious systems, but I am not sure if a RH can be used in accepting the basic religion. It may be that the few propositions its creed contains would all be accepted directly by intuition. That is, anyway, how Dilthey conceived the acceptance of a *Weltanschauung*. It is, I think, a possible, but not very probable explanation because it seems that most people accepting a basic religion do so under the influence of other people, i.e. are using authority. But if so, then an inference is present and it is plausible that the conclusion of that inference is a RH.

III.

The second important correction which I would like to make in my earlier views is concerned with religious experience.

The curious fact about my book is that although it is about religion, it does not contain a single mention of religious experience. I did not say so explicitly, but I seem to have assumed that there is no religious experience at all. This did astonish, not to say scandalize, several religious thinkers. It can be psychologically explained, among other reasons, by the fact that I came to the study of religion from that of Marxism-Leninism and simply projected into religion what I found in the latter.

Anyway, it was a Marxist-Leninist who pointed out to me, that my theory of RH does not only apply to religion, but also to his own beliefs and to similar *Weltanschauungen* (in the modern meaning of the term), like those of Illuminism and National Socialism.

And that is also one reason for the weakness of that theory – it does not take into consideration one quite essential specific feature of religion: religious experience. Bertrand Russell committed, I may say, the same blunder when he wrote that Communism is a religion of the Islamic type. Perhaps the reason was the same for both him and myself – an exaggerated rationalism.

I now think that it is necessary to admit the existence of religious experience but that a twofold distinction should be drawn in the meaning of the term. One should distinguish a religious experience proper from the religious explanation or interpretation of secular experiences – and, in the religious experience proper, between the encounter with the Transcendent and the meeting of the sacred in the world:

the sacred is
- (a) introduced by explanation
- experienced
  - (b) in the encounter with the Transcendent
  - (c) in the world

In the book I admitted that the sacred is introduced by the RH (a), but claimed that the experiences it explains are secular, i.e. I rejected (b) and (c). Actually I am enclined to admit (c) and, with qualification, also (b).

As to (b) many writers say today, that there is a sort of communication, or dialogue with God and consequently that the believer experiences the sacred in Him. But I do not believe that. It is true that Bergson contended with convincing reasons that some exceptional believers, mystics, like St. Catherine of Siena, did meet God. But I do not have any similar experience myself and suspect that neither does the majority of other believers. When they say they are meeting God, they are using the term "meeting" in a peculiar, I would say abusive and misleading, way.

But as to (c) I think that, given the authority of the great majority of scholars who studied religion, it is necessary to admit that there is some peculiar religious experience, conceived as meeting the sacred in the world. This does not mean, by the way, that anything real, in particular that Transcendence necessarily corresponds to that experience. A nominalistic explanation (as professor Hick calls it) is possible.

But if a realistic explanation is accepted, then the RH, while remaining an explanation of the whole of a believer's other experiences, is, first and above all, a hypothesis explaining his religious experience.

Some difficult problems arise here.

(a) One is this: should we say that the sacred is a specific value which cannot be reduced to other values, or not? If not, the religious experience would be just a complex of secular experiences. The authority of many prominent thinkers is in favour of the specificity – to mention only Scheler and Otto. But at least one great thinker told me once, he could never understand what the sacred is. So I must say I do not know the answer.

(b) If we follow, however, the majority of the philosophers, and accept the specificity of the sacred, the further question arises – where do religious people meet with it? It may be that they encounter it in nature – think about Einstein's "oceanic feeling" – or in saints. Again, I do not know the answer to that question.

IV.

The hypothesis that people experience the sacred in their encounter with saints is relevant for still another correction. I said in my book of 1965 that an intellectual cannot use authority for the justification of his religious faith. That now appears to me to be incorrect.

I was led to my present view by studies in the Italian Renaissance and also by some considerations of Ciankara's success in India. In the late XVth century in Italy we have the case of Pico della Mirandola, perhaps one of the most brilliant minds of his time, the pupil and spiritual son of Marsiglio Ficino. Now Pico had been so much converted to the peculiar brand of Christianity preached by Savonarola, that he let himself be buried in San Marco in a Dominican habit.

And I learned this about Ciankara: this great man did personally convert a great many people to Hinduism, and among them there must have been many intellectuals of high standing, given that so many among them were Buddhists.

In both cases such intellectuals seem to have justified their acceptance of a faith by authority.

Therefore it seems to me now that two cases must be distinguished here according to two sorts of authority. Sometimes the authority concerned is an individual prophet, preacher etc., in other cases it is a community. Both sorts are frequently exemplified in history.

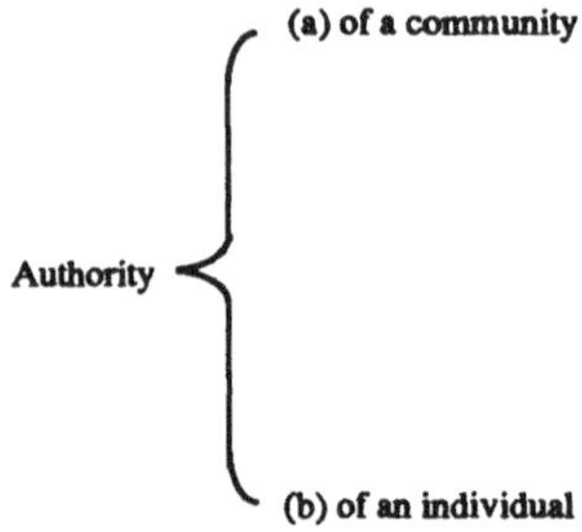

Now what I said about the argument of authority not being available to intellectuals seems to apply only to the authority of a community, (a), not to that of individual saintly preachers (b). The authority of an individual like Ciankara or Savonarola plays, on the contrary, a considerable role in the conversion of some, sometimes prominent, intellectuals.

The consequence for the theory of the RH is a limitation of its scope. The RH will still be the justification for the faith of many intellectuals, but not of all of them.

V.

A fourth problem concerning the RH arises in the following way. Contrarly to what I said in the book, there seems to be still another difference between the scientific and the RH. The former allows for forecasts, while it seems to me that the RH does not. Thus it has been suggested that the RH, lacking an essential feature of every hypothesis, is not a hypothesis at all.

The problem seems to be partly linguistic – it is a question to know what we should call "hypothesis". Those who say that every hypothesis allows for forecasting clearly assume that the term must be used in the way in which it is used in natural sciences. But this is not the only use of the term. Both in ordinary language and in mathematical logic it denotes sometimes, so it seems, sentences which do not pretend to allow for predictions. The only feature common to all those uses of the term is that it refers to a sentence which explains, i.e., orders propositions of a certain field. And the RH certainly does that.

It may be further objected that a sentence which does not allow for predictions has no practical value. But while it is true that the RH is not a guide for research, nevertheless it has a high practical value, as it always includes a code of behaviour.

Those answers would, of course, not be accepted by somebody who takes all human knowledge and rationality to be of a unique type: that of the natural sciences. That was, in fact, an assumption made by the Enlightenment. But there is no reason to accept that view today, and there are many reasons to consider it obsolete.

***

To sum up:

1) I now think that the theory of the RH has to be corrected in three ways:

- as to its function: the RH will find an application above all in regard to other religious systems; whether it can also be used also to justify the acceptance of basic religion is doubtful;

- as to its foundation: it should also explain religious experience;

- as to its users: even intellectuals may use other ways of justifying their religious faith.

2) But given these corrections and some explanations it remains an acceptable theory - the only acceptable one for some cases.

3) Yet much research is still needed in the field. Up until now logicians and analytical philosophers of religion have mostly concentrated on semantic problems; little work has been done about questions concerning the structure and the justification of faiths. Such studies would be welcome.

DISCUSSION

WEINGARTNER: I'd like to mention a view concerning religious experience, which I found in Aquinas. His opininion was roughly that first of all we have to admit that some persons have religious experience, i.e. that such a thing is possible. There is the story that at the end of his life Aquinas also might have had such an experience. But then the second question, according to Aquinas, is, whether religious experience can play any role in theology in a kind of scientific attempt about religion, and here he was quite strict to say no. It cannot, because firstly it cannot be communicated to others. Secondly, others would not understand this kind of personal and individual experience. It cannot be communicated to others, because all those people who have it, cannot even describe it to others, neither in scientific language nor even in everyday language. The experience is bound to the person and individually specific to such a degree, that it cannot be described for others. Therefore we cannot use religious personal experience as a premise in any theological argumentation. But communication is a necessary condition for science according to Thomas.

BOCHENSKI: If I understood you correctly, according to Aquinas the use of religious experience in the preambula fidei is a justification for faith, but not in theology. I am doing some work on the first *question* of the Summa Theologica right now and though it is the work of a genius there are passages which I find quite confusing. About the treatise of De Deo Uno it was so easy to give theological arguments, there is not a single argument from the Scripture. Everything is purely philosophical. Why? It may seem, because his view is the deductivistic one. That you must first prove the existence of God an then so on, the existence of Christ etc. But I don't think that is his view. But what then is his view? Why did he build in the questions 2 to 26 such a deductive system? I think, because it is nice to have such a complete system, I cannot see any other reason, because it is not theological and not a preambula fidei.

WEINGARTNER: You said, and I think, this is an important point, that the religious hypothesis tries to explain everything or is a kind of super-explanation. Hempel has written a short essay about the question whether such super explanations can be used in science. According to his view we cannot, i.e. in science we have to stop somewhere before. Now this is certainly a tendency of religious hypotheses, I agree with this and certainly there is a difference between religious and scientific hypotheses. It is also my view, that religious hypotheses also try to explain values, but here I do not see such a big difference as you, because I accept, that there are also scientific disciplines which at least see it as their task to explain values, that means, to give reasons for values. In physics and in mathematics this occurs only in the methodology – one searches for good means for a presupposed aim, but in ethics and pedagogics also reasons and arguments for the aims are investigated. Even if one says that there is a distinction between an empirical pedagogics and perhaps a non-empirical one, I think, it is an amputation of the discipline to take only the first and not to discuss the aims of education. Also the theory of literature is evaluative and not only in the sense that it presupposes values, but also in the sense, that it really tries to explain and to give reasons for evaluations. A further important point was your question whether religious hypotheses make predictions. I think yes. Concerning predictions we have to distinguish two things, predictions inside the world and predictions outside the world or "after" your life in this world. According to Thomas Aquinas one cannot prove in this world, that it has an end in time and so the belief in this is a matter of faith. So there is the prediction, that this world will sometimes collapse. There is also the prediction of the destruction of Jerusalem, but there was a discussion in the second century, whether this was a real prediction or done "post festum". Thirdly there is the prediction in religion that life or consciousness goes in some way on after death. If somebody says, this is not an empirical prediction, one could answer: wait! You will find out yourself. So it is not an empirical prediction in the sense of being intersubjectively provable but in the sense that everybody will experience (find out) for himself.

BOCHENSKI: I think, I understand your second and third point, but I am not quite clear about what you said about generality. You see, I think, a religious hypothesis explains the totality of experience and brings an order into it.

WEINGARTNER: Yes. On this I agree with you and I wanted to stress that this is something which we do not have with scientific hypothesis, even it is suspect, if a scientific hypothesis tries to explain everything, because we cannot test something which explains everything.

BOCHENSKI: You see, I tried hard to understand something what people mean when they speak of the poetical meaning of life. It is hard to say what they do mean really. Well, I found out, that this is ordering of the whole experience. But I think, I am in disagreement about your second point. You said, if I understand you correctly, that even in science some value-proposition may be stated, is it correct?

WEINGARTNER: I do not say in natural science except in the methodology, because even in natural science and in mathematics there is an establishment of methodological rules.

BOCHENSKI: But they are hypothetical!

WEINGARTNER: Yes, I agree, because we do not want to change the aim of science, which is truth or comprehensive truth. Take for instance the rule: 'The new hypothesis should include the correct results of the forerunner hypothesis'. Physicists added sometimes, it should not only include the forerunner hypotheses, but it should include them as special cases. If this rule is violated, there is danger, that you have false claims or false consequences in your new hypothesis and this is again incompatible with the aim (truth). I agree with you, that this rule is of course connected with the aim (if the aim should be reached then the rule should be applied) and in this sense is hypothetical. But such normative rules and principles can also be revised and refuted as the principle of continuity shows. It said: "Look for continuous dependencies of causes and effects", and was a long time a very important rule. Then by quantum mechanics it was revised or better to say restricted or refuted, so there *is* also revision and refutation in the methodology, but of course, with the help of aims. A similar situation holds in the methodologies of all sciences. The situation is different however if we talk about teleology. Teleological explanation presupposes values in the premises but not in the conclusion as I see it. So it explains facts with the help of values for instance, a historical fact with the help of motives and aims of persons. So values are here presupposed, not established. But literature, in my view, is a clear example, where value statements are established. There value statements (containing aesthetic values) are given some justification.

BOCHENSKI: Well, this is a point of disagreement. I think, the point where you can best exemplify these problems is morals. There are three completely different fields: it is first a class of rules, categorical imperatives. Then second there is, what you call "metaethics". (I call that "ethics".) You ask, why there exists such a rule or what is conscience and so on. And third, there is the technology of good life. It is a technology, like that you were talking about: If you wish to obtain good life, do this and that. I confess, that I have written a book of Wisdom, where the principle is: Behave in such a way, that you may live long and in a pleasurable way. There are many technical imperatives in sciences, but I cannot see how a scientist could state value propositions, because sciences are interested in what *is* in the world. It is a presupposition, that technology is good. But this is a presupposition, not a statement, say of physics, of astrophysics or of biology.

And there, I think, is a radical difference between a scientific hypothesis and a religious hypothesis.

HICK: I'd like to comment on two, in particular, of Prof. Bocheński's main points. The first is the distinction between basic religion and specific religious systems. It seems to me that some distinction of that sort is essential, if the philosophy of religion is to be more than the philosophy of one's own specific religion; and most philosophy of religion still is simply the philosophy of one's own religion. But if you are going to take account of religious systems generally then, I think, this kind of distinction is essential. By particular religious systems we would mean Christianity, or more specificly Catholicism or other branches of Christianity, or on the other hand Buddhism, Islam, Judaism, Hinduism, and so on. And by basic religion I think we mean that which all these systems have in common, and what they have in common, in very broad terms, is a reference to the Transcendent, or Transcendence, or Reality, or the Real. One virtue of the term 'the Real' is that it corresponds pretty closely to the Sanscrit sat and also to the Arabic al Haq. Now, if we do distinguish between basic or, if you like, generic religion, and specific religions, then we also have to distinguish between their intentional objects or focuses of concern. The focus of concern of Christianity is the Holy Trinity, of Islam Allah, of advaitic Hinduism the non-personal Brahman, etc. And the focus of interest of basic religion is the Transcendent, the Real as such. But then, of course, the question is going to arise: What is the relation between the Transcendent and the specifically conceived Holy Trinity, and Allah, and so on? It seems to me that the answer is to be found along these lines: the specific objects of worship, the Holy Trinity, Allah, Vishnu, Shiva, etc., represent different human perceptions of the Transcendent.

Whenever we experience anything, there is an important human element in our experiencing. We are using a system of concepts, and in the different religious systems different sets of religious concepts are being used, with the result that the transcendent reality is differently conceived and therefore differently experienced, and therefore differently responded, too. So I want to welcome Bocheński's distinction. I think it would make a very big difference in future work in the philosophy of religion if some distinction of that sort were adopted.

And secondly, I'd like to agree with Paul Weingartner about prediction. I think, prediction is too narrow a word: but the religious hypothesis in its basic or generic form, as also in its specific forms, offers a conception of the nature or structure of the universe as a totality. We can compare the religious hypothesis with the naturalistic hypothesis, which is its polar opposite. They offer pictures of the universe, such that the future development of human experience will be different according as one or the other hypotheses turns out to be true. Of course the major distinction is in some kind of belief in immortality versus the belief that human life terminates totally at death. If it terminates totally at death, then of course there will be no one to notice that the naturalistic hypothesis has been confirmed. But if on the other hand the religious hypothesis is correct, then we shall have experiences which will indicate to us that it is correct in contrast to the naturalistic hypothesis. So it does seem to me that in this very basic way it is not totally dissimilar from a scientific hypothesis.

BOCHENSKI: Thank you. Now I would like to make the following remark: You say, and this is new to me, you think, that the basic religion is common to the book-religions. I forgot to say in my paper, that what I call religion, is the class of the great book-religions, Brahmanism, Buddhism, Mosaism, Christianity, Islam etc. There are so many different religious phenomena that I always thought a definition of religion in general can hardly be given. Now do you suggest, that basic religion would be common to those five? That's a question. I am not so sure of it. It may be, that even basic religion is different in different religions. It reaches the transcendent in different ways. I would ask you as an authority in the field.

That's one thing. Now, concerning the end of life, you think, that something essential to basic religion, that life does not terminate with death.

HICK: This is certainly common to all of these religions you have mentioned – in different forms, of course.

BOCHENSKI: I am not sure of it. There is a famous case of Unterlinden, a Convent of Dominican Sisters in Elsaß, who led a very severe and mortified life. A delegate of the Master of the order was asking once (it is a historical fact) the prioress, what would they do, if they would be told that there is no survival. And the prioress said with great astonishment, 'But Father, obviously the same thing we are doing now'. You see, that's another fact. There is also this sort of religion, which is not bound with that personal survival.

HICK: Yes, it's true.

SCHURZ: I have a remark and question concerning religious experience. It has been said, that religious experience has some peculiar features, different from other experiences, for example it is secret, you can't communicate about it, it is transcendent, and so on. But I don't see reasons why it should be secret or why I could not communicate about it, if I would have it. There are so many experiences in our lives, which are extremely intimate, dreams, other things, etc. and we communicate about them all the time. I think, if I would have religious experiences, there would be no problem for me to communicate about it.

BOCHENSKI: You think there is nothing specific in religious experience?

SCHURZ: No, there is no reason to assume that there is something specific, different from dreams, from other psychological experiences, no real reason, that it could not be intersubjective in principle.

BOCHENSKI: There are two different points. Most of continental philosophers, who were occupied with religion, were asserting the specificity of the religious values and the value of a second noumenon. Take the theory of Scheler: you have got different classes of values. The values of the pleasant, the values of writer, the spiritual values and then at the top there are the religious values. For instance the value of the victory is a completely different value than that of the beauty. So is religious experience experience of something quite different according to Scheler from the experience of other values. You say, that you don't see its specificity. You are not the only one. I quoted Ingarden, who told me, he could never understand what it is, you see. To someone from the other school we say, 'Well, they are simply blind people'. With the values it is like that. In the case of the Nazi-Jew well, many were motivated to sacrifice themselves for something which is essentially negative to values. So the Nazi-Jew certainly had some positive values, by which they were convinced. What were the point obviously general language – the values of the Bible! Not the values of the pleasant, nor the spiritual values or religious values. So it happens. I may speak about myself, very personal. I don't understand anything in music, anything. It is rather strange to say that in Salzburg my parents forced me to make 6 years of piano. Then I came to a very great technic. But I am completely blind of musical values. So those people, who believe in those values, will rather say that you are blind in respect to them. And I think that the great majority will be against your position.

SCHURZ: It is a kind of immunization. Look, what are your reasons for believing, that these kinds of experiences are secret or something private? Tell me the features of them, and then I will show you that other experiences have them too. You are speaking about music; what about experience in love, it has always properties, you can't describe it, you really can't say it, but you can communicate with other people about it. So I think, there is no real reason, why you should not be able to communicate about a religious experience, if it would exist.

BOCHENSKI: There are no possible arguments, either you see or you don't see. And so some people are blind of music, like me. You can't explain it to me, what the greatness of Mozart is.

SCHURZ: Think about music. Assume I speak with someone about that beauty of Mozart and he tells me, 'There is no argument – either you hear it or you don't hear it, I can't explain it to you'. Assume I speak about music, about say Mozart, yes, I want to learn something about Mozart, I want to learn something about music, and the artist tells me, 'Well, there is no argument and nothing to learn, either you hear it or you don't hear it'. So, that would not be very reasonable in art. So why should it be reasonable in religion?

BOCHENSKI: I am not reasoning about this experience, I am not denying that you are right – but you see, it is your position. I proceed in my book, with the presupposition that the common believer has no religious experience, but a religious explanation of his other experiences. There is, however, a powerful authority of the great majority of specialists, both in history of religion and in philosophy of religion, against your position, not against mine, because I am rather neutral in that point. Now, where I am not neutral, is, when you try to make arguments, to prove that there is religious experience. I think, you cannot prove this.

HICK: It seems to me that the idea that religious experience is totally indescribable is very strange, because religious literature is full of descriptions and discussions and communications of both of the kinds that Prof. Bocheński distinguished, that's to say specific encounter experiences and also the more generalized type in which events in the world are experienced as having religious significance. And they are no more ineffable or indescribable, or incommunicable, than many other kinds of experience. So if we start out by saying that religious experience cannot be communicated or talked about at all we are going to make life very difficult for ourselves, and unnecessarily so.

WEINGARTNER: I want to continue here. I think, what you say now, John, is already in the direction, that one has to distinguish different levels or categories in religious experience. Because the extreme side is this kind of experience, where we say, this is hardly to be accessible or communicable, this kind of thing, which was described at least in some sense by the mystics like Tauler or Saints or others, who said, they had such an experience, if they described it at all. And here I think, I would personally agree with Bocheński, that this is probably inaccessible: you have this kind of very personal thing or you don't have it. And similar things you can even see with children – as long as they don't have a certain kind of experience, then you can try to explain a long time, it does not help at all! (For instance each of my children wanted to try to put the fingers on the hot stove and burned them to make the experience of 'hot'). The second is a much weaker kind of religious experience. This is, what religious persons sometimes say, that they have. I mention now a perhaps controversial view of a theologian at the University of Salzburg, Prof. Beilner. He has the following theory: He says, if you try to obey the moral rules of the Bible, then you will have a kind of empirical experience, that you feel better, you feel free of guilt, you feel happy in some sense. There are such predictions in the Bible (St. Paul), which say, that if you live the "spiritual life", your reward will be love, enjoyment, peace, friendship, patience, goodness etc., but if you live the "life of the body", then the consequences will be hate, anger, malice, greed, lewdness, calumny ... etc. And Beilner says, that for him this is a kind of religious experience where we can even test these predictions. One is invited as a religious person to try it. And this level of religious experience is certainly communicable. I think, it is similar with those examples mentioned by Schurz. There are also different levels. Although 'falling in love' is certainly communicable, what great and deep love is cannot be explained to a twen, who never had such an intensive experience.

KUTSCHERA: If you don't have a certain experience, you can't deny, that there is such an experience. If you have no experience with music, you would not normally deny that there are musical experiences, or generally: aesthetic experiences. Now people who talk about them understand each other, they discuss the aesthetic qualities of a work of art in a meaningful way. So there evidently is aesthetical experience. People also talk about their religious experiences. The only problem is, that if you don't have this experience, then it can of course not be ground for your belief.

BOCHENSKI: That's the point. Well – you see, in my book there was the claim that there is no religious experience in common believers. Now I am not so sure about it. – But in the book it was my opinion, that you can have religious hypothesis without having a religious experience of the sort I was talking about, namely the meeting of the Holy in the saints. It is even possible that people do meet it in some secret way in nature, like in Einstein's oceanic feeling. What I denied in my book, is that common believers (as opposed to mystics) do meet God. I think, this is certainly not the case.

SCHURZ: I think it's important to understand religious experience even if you don't have it, because the important question is: "to what degree is the assumption of God necessary to explain religious experience?" And this question is a question for everyone. So what is important for me is the possibility to understand religious experience even if I don't have it. And I am asking myself, to what degree it is possible, and I think, it should be. One example is a famous example of the psychologists and psychiatrists. A psychiatrist does not need to be schizophrenic, to be able to understand the schizophrenic. To understand a depressive one, you need not be depressive yourself, luckily. And so, I think, it is not necessary for me to have religious experience to be able to understand if the other person is really willing to tell me, if he is not just isolating himself. The examples of Weingartner are certainly correct, but this is a special case: if you are speaking about adults, all capacities are developed, the mental as well as the sensorial ones. I mean, there are good reasons, that a child cannot understand all things, but if we are speaking about adults, we have similar capacities.

KLEVAKINA-ULJANOV: My remark has to do with the idea to what extent it is possible to comprehend religious experience of others never having it oneself. As Professor Schurz, I am not inclined to treat personal religious experience as a decisive factor for understanding the discussed phenomenon, but I would not also ignore the role of religious "sympathy" (i.e. potential readiness or open-mindedness to accept religious claims) in this respect. My point is that in order to be able to grasp the state qualified by an individual as a religious experience, one should be prepared to intuitive apprehension of this state in general. In my view, this could happen only if one is "sympathetic" to a general line of thought of a religiously oriented person. For example, there are a good many atheists who would deny the possibility for a man to stand in a relation to whatever one may consider the divine. Again, people who are religiously inspired, or at least those who are not radical materialists, would disagree. So my view comes to this: religious sympathy is not the same thing as religious experience; but without the first the second would not come into existence and what is more, it would not be open for an outward comprehension.

BOCHENSKI: There is something you should have said, that according to the Christian view we live in the darkness of the faith. We believe only what we get from the scripture, from something written on paper and not from another experience. Well, this is my interpretation of St. Paul. So it is not so sure, that religious belief is linked with religious experiences, it maybe, the majority of specialists say it is, but I don't know. There may be different possibilities. One possibility is that which is suggested by St. Paul, that is, that there is no religious experience. We have got only the secular experience and in order to explain that you assume that religious hypothesis which raises up to the Transcendent. That's one position. The other possible position is, that alongside with the set of secular experiences there is also some peculiar religious experience. And then the third possibility is, that you have got an encounter with the transcendent itself, with God or whatever it may be, the Absolute. Now I repeat the suggestion, that the Christian view is more akin to the first hypothesis. There is here no religious experience, we live in the darkness of the faith, we accept a set of revealed propositions.

HICK: Just in passing and in brackets, concerning St. Paul. His Christian life began with a momentous religious experience on the Damaskus road. But it seems to me that there is a fundamental division between two approaches to the nature and validation of religious belief. One is headed as its great figure by St. Thomas, namely it's all a matter of philosophy, it's a matter of using your reason to find the most probable explanation of the visible world. I would say that that does not lead to religion at all. It may lead to certain philosophical views, such as the view that there is a God. But this is not in itself a religious view, in my opinion. It's only that if one responds to the thought that there is a God in a certain very distinctive way. On the alternative view religion is not a theory, a philosophical concept, it is an actual response, which has many dimensions including emotion, to the Transcendent or to the mystery of the universe. To my mind the project of trying to prove the existence of God, or trying to show that it is more probable than not, has totally failed. This is a controversial saying, but if it is correct the alternative approach to religion becomes all more important.

BOCHENSKI: I think, that what you say is quite correct. Then, if you think about the secular experiences and then make the hypothesis, that there is a God, that is Natural Theology. That is philosophy and my experience of history of human thought makes me believe, that this is probably the most popular conclusion among philosophers. If you would ask me, what is most popular among philosophers – we take all the history – it is the affirmation that there is some Absolute, a sort of cornerstone for rationalism. Perhaps not true for our time, but during whole history.

But that is not what I meant by the religious hypothesis. The hypothesis in this case is not God, but the providence. Not the God of Spinoza, let's say, or it is the matter of Dialectical Materialism. But God, who is a good father. That I think is properly a religious thing, and that makes the distinction between the God of philosophers and that of religion. There is a bitter statement of Whitehead, that probably nobody went farther, purely philosophically, in natural theology, farther without some influence of religion, than Aristotle. But philosophical inside did not lead him very far to produce something available for religious uses. The God of philosophers is certainly not the God which religious people want. I am personally very much opposed to people, who confuse demonstration of the existence of God with religious hypothesis. But what I think is essential to basic religion is that assumption, that there is not only an Absolute, but (at least in our culture) that there is providence.

JOHN HICK

ON RELIGIOUS EXPERIENCE

If we start from ordinary usage, we can say that interpretation is concerned with meaning, and presupposes that there is something (using that term in its most comprehensive sense to include entities, statements, actions, complex situations, or indeed the universe as a whole) whose meaning is not indisputably self-evident to us. There is accordingly ambiguity, making room for alternative construals, some of which will normally be misconstruals.

Ordinary usage thus suggests a dichotomy between objective facts (statements, actions, entities, situations, the universe as a whole) and subjective interpretations of them. However at this point we need to distinguish between the two main families of meanings of 'meaning': on the one hand the various kinds of semantic meaning (i.e. the meaning of linguistic utterances) and on the other hand the kinds of what I shall call dispositional meaning (i.e. the practical meaning, for the interpreter, of objects, events and situations). In the case of semantic meaning there is indeed a dichotomy between a linguistic entity – a sentence, an exclamation, a command, etc., – and an interpretation of it. But dispositional meaning is importantly different. The world is indeed there, and is as it is; but we do not have access to it as it is in itself, unperceived by us. We are aware of it only as it impinges upon us and is perceived and inhabited by us in terms of many kinds and levels of dispositional meaning. The dispositional meaning of an object, event or situation consists in the practical difference that it makes, currently and/or potentially, to the meaning-perceiver. For example, I perceive what is before me as an orange. In so doing I am recognizing or identifying something by means of the concept 'orange'. And my recognizing it as an orange consists in part in my being in a dispositional state in relation to it which is appropriate (as I take it) to its being an orange rather than something else. Such a dispositional state usually cannot be fully spelled out. But it includes in this case being liable in certain circumstances to eat the orange; and it excludes, for example, expecting it to talk or grow wings or prove to be as heavy as lead. Thus when I see or, using all the relevant senses together, when I experience this as an orange my total dispositional state includes a sub-range of dispositions that is appropriate to this thing being an orange. And the same holds for everything else that I recognize, i.e., am aware of as being some particular kind of thing or, in other words, as instantiating some concept.

The dispositional meaning of events is more complex than that of individual objects, being usually an aspect of the yet more complex and comprehensive significance of a situation. A situation is composed of objects but has its

P. Weingartner (ed.), Scientific and Religious Belief, 161–182

own dispositional meaning over and above the separate meanings of its constituent objects. Consider for example the situation which we describe as a session of an academic conference. The participants, having been prepared by invitations, programs and other documents, are automatically experiencing what is going on around them as a session within, let us say, an American Philosophical Association meeting. They are in a dispositional state to behave appropriately – by listening to the paper, being ready to raise questions and to discuss after the paper has been read, and so on. And this rather complex readiness to behave in certain kinds of ways and not in others presupposes an extensive network of concepts which are part of our modern western academic culture. But if we can imagine a stone age person suddenly being brought into an academic conference, such a person would not perceive what is going on as having the same character or meaning. The stone age person would not have the concepts of conference session, academic discussion, scholarly paper, philosophy, or most of the wider conceptual field to which these belong, and would accordingly experience the same physical configuration as having some quite different meaning to which a quite different dispositional state would be appropriate.

Human life is normally lived at this situational level of complexity. And whereas the dispositional meaning for us of natural objects, such as oranges, consists in our practical adaptation to the physical world, situational meaning is largely a cultural construct. For our inhabited world of meaning – corresponding to the Lebenswelt of the phenomenologists – depends upon our corporate systems of concepts, which have formed over the decades, centuries, or even millennia, and are embodied in our developing languages. Our experienced and inhabited world accordingly varies considerably through time and across cultures.

Returning now to our imagined conference, its participants, although intend upon the business before them, are at the same time potentially within a range of other situations. On one level (in a sense of 'level' to be indicated in a moment) they are not only at the gathering in, let us say, San Francisco, but also in the larger situation of the life of that city and the still wider situation of the earthquake region of California. And if an earthquake were suddenly to occur their situational awareness would at once shift to this new context of meaning, evoking a different dispositional state, expressing itself in appropriately different patterns of behavior.

In such a case the new focus of attention would supercede the previous one. And it is a feature of situations of this kind, on what I am calling the same level, that they are more or less mutually exclusive, so that we can usually only live effectively in one at a time. But we can, and often do, live simultaneously in relation to different levels of meaning – levels in the sense that a higher includes but transcends a lower level. The three such levels that have long been recognized in western thought are the physical, the ethical, and the religious. Notice that in each case meaning has an essential behavioral aspect: the meaning of an object, event, or situation is defined in terms of the practical dispositional state in which one is in virtue of recognizing that object, event, or situation as having that particular character. ('Recognize' here of course includes 'misrecognize', for when we make mistakes in identifying what is before us or around us we still respond by varying our dispositional state in relation to it, though this is then an inappropriate instead of an appropriate variation).

On the physical level of meaning are all the ways in which we experience material objects, events, and situations as being such that it is appropriate for us to behave in relation to them, or within them, in one range of ways rather than another. To experience this as a pen, that as a table, and that over there as a mountain, and to experience a conference, a family meal, walking on a sidewalk, driving a car, and so on, as we do, are examples of our continuous awareness of meaning at this basic level. It is because we perceive our environment as having meaning in this dispositional sense that we are able to live from moment to moment and year to year within it.

Such a dispositional analysis of recognition in terms of concepts presupposes of course some basic aim or desire on our part, such that this rather than some other set of readinesses for action becomes appropriate. Roughly, in relation to our physical environment, the basic aim that renders a particular dispositional response appropriate is the aim of surviving and, beyond this, of flourishing in terms of such basic natural values as health and contentment.

Our ordinary everyday consciousness of our physical environment is thus normally a continuous consciousness of it as having various kinds of meaning in virtue of which we have some notion of how to behave within it in order to fulfil our basic aim as animal organisms. This everyday consciousness normally occurs at the situational, rather than the object, order of complexity. All this applies to the lower as well as to the human forms of animal life, though with two differences: the situations of which we humans are conscious can have a much greater temporal extension; and what are in the lower animals simply recognitional capacities have been abstracted by the human mind as concepts fixed by language.

It is within our human situational awareness that the next level of meaning, the ethical, arises. I shall say very little about this here, since it is not the primary concern of this paper. But situations of which other persons are constituents always have actual or potential ethical meaning for us. That is to say, they may render appropriate some action or range of actions determined, not by the aim of surviving, but by a moral principle. In my view Kant successfully identified the basic nature of the distinctively moral aim as that of treating oneself and others equally as ends in ourselves. In terms of this basic aim, interpersonal situations can take on a further meaning, whose appropriate dispositional response may be at variance with, and claim to override, that rendered appropriate by their purely physical meaning. Thus anyone who incurs death or pain or even discomfort or inconvenience in order to treat others as ends rather than as means is responding to the distinctively ethical character of some situation. The physical situation remains unchanged, but the way in which it is experienced has changed, in that it is now perceived as also having moral significance.

Now let us turn to the religious level or order of meaning. Here we apply religious concepts both to external objects, events, and situations and to the events and states of our inner life. On the face of it some forms of religion are more concerned with the inner and others with the outer realm. However we must not make the mistake of characterizing whole traditions in these terms – for example, as exclusively 'prophetic' or exclusively 'mystical'. For the various kinds of religious experience constitute, I shall argue, a continuum ranging from the purely 'external' to the purely 'internal', the entire spectrum being present in varying proportions within each large and long-lived tradition. Further, I shall argue that religious experience throughout this continuum consists – if understood religiously – in the presence of a transcendent divine Reality coming to consciousness in terms of our human concepts. Because the different ways of being human have produced a variety of such conceptual systems, with their associated spiritual practices, the transcendent Reality (which I shall refer to simply as the Real) postulated by a religious understanding of religion is experienced in a variety of ways which have become enshrined in the different religious traditions.

I have spoken of the presence of the Real coming to consciousness in us as our awareness of the Real. In the case of physical realities, their presence to us comes to consciousness by means of their impact upon our sense organs, which is somehow translated into modifications of consciousness that are endowed with dispositional meaning by the system of physical-object and physical-situation concepts in terms of which we live. In contrast to this the presence of the Real affects us by a non-physical impact, which comes to consciousness in forms that are endowed with dispositional meanings supplied by our system of religious concepts.

This occurs in a range of ways. At one end of the spectrum – where most moments of religious awareness are concentrated – our attention is directed upon the material environment, but we experience this not only in terms of physical concepts but at the same time on another level, in terms of religious concepts, so that some object or event or situation is experienced as mediating the divine presence to us. At the other end of the spectrum the divine presence is experienced independently of the physical environment, in forms supplied by the mystic's religious concepts. Between these extremes are various mixed modes, as I shall indicate in a moment.

Let us note some examples along this spectrum. The kinds of objects that are experienced as having religious significance include icons, images and idols (such as are found in Orthodox and Catholic churches and Hindu temples), symbolic objects (such as a cross or the star of David or the bread and wine in the Eucharist), and also holy places (such as Mecca, Jerusalem, Bethlehem, Vrindaban, Benares, Bodh Gaya). To experience these as having religious significance is to be in a dispositional state in relation to them that can be broadly characterized as reverence, a receptivity through which (on a religious interpretation of religion) the Transcendent is mediated to us.

However religious experiencing-as more commonly occurs in the awareness of situational than of object-meaning. Thus at the situational level the religious consciousness may find a further order of meaning in the moral life, interpreting the ethical requirements of the interpersonal world as mediating either the external claim of God or the internal requirement of Dharma, leading one to act or refrain from acting in this or that way. But it may also find a religious meaning in any or all of the events of individual life and corporate history. Thus in Hebrew religious consciousness it has embraced the entire national story, seen as a living out of the covenant with Yahweh. For the understanding of history expressed in the Hebrew scriptures is one in which crucial events have a religious significance. The exodus from Egypt, the wandering in the desert, the settlement in Canaan, the exile in Babylon, the return, are all presented as happenings in and through which God was guiding, disciplining or caring for God's chosen people. Much of the story, as it is told in the Hebrew Bible, has reached mythic proportions, and we do not know precisely to what extent it is rooted in history. But in some of the prophetic writings we come closer to history as it was taking place and was experienced at the time as having religious significance. For example, it seems that Jeremiah in the sixth century B.C.E. experienced the Babylonian army marching on Jerusalem as being wielded by Jahweh to punish faithless Judah.

It is in such historical events that we find the foundational religious experiences that have given rise to the three Abrahamic religions of the Book. Each is based upon special revelatory moments in which historical events were experienced as divine epiphanies. Thus Moses' experience at the 'burning bush', and in receiving the Law on Mount Sinai, and again the communal experience of the exodus from Egypt, were foundationally revelatory events for Judaism. Jesus' intense experience of God as abba, Father, expressed in his life, teaching and death, including the unknown event that has come to be called his resurrection, is the foundationally revelatory event for Christianity. And the prophet Muhammad's experience, over a period of some twenty years, of hearing a heavenly voice reciting the words that became the Qur'an, is the foundationally revelatory event for Islam.

Each of these required an appropriate interpretation, or mode of experiencing-as, in order to have the character of revelation. Considered simply as historical events each is capable of being construed both religiously and naturalistically. When someone – Moses, Jesus, Muhammad – reports that they were conscious of a divine presence or heard a divine voice, it is always open to the sceptic to grant that they had the experiences that they report but to hold nevertheless that these were hallucinatory in character, being projections of their own unconscious minds. Thus sceptics have held that Moses was deluded; that Jesus' intense communion with God was a religious hallucination; and that the suras of the Qur'an are purely the product of Muhammad's own mind, without any transcendent input. We are noticing here an aspect of what we can call the religious ambiguity of the universe – the fact that, from our present standpoint within it, the universe is capable of being intellectually understood and concretely experienced in both religious and naturalistic ways. I shall come later to the question whether, and if so how, this ambiguity may be resolved. But first let us look more widely at the religious way of experiencing what is happening around us and within us. Within the theistic traditions the broadest term for the dispositional aspect of this mode of experiencing-as is worship. But this includes much more than specific acts of prayer and liturgical behavior. It includes a centering of one's existence in God, expressed in obedience and trust and in a consequent release from self-centeredness and a freedom to love others; and also in an ultimate confidence that (in Lady Julian's words) 'all shall be well, and all shall be well, and all manner of thing shall be well'.

To depict in this way a life lived in the conscious relationship to God that we call worship is, of course, to picture an ideal. Most of us live most of the time in forgetfulness of God and with our existence centered in ourselves.

Nevertheless this God-centered form of life remains the ideal which we see embodied in considerable degree in some of the great saints of our different traditions, and in many much lesser degrees in many ordinary believers. When such a believer reports the 'sense of the presence of God', or perhaps better the sense of being in God's presence, she is experiencing her total situation as mediating the divine reality. This mode of experiencing-as may be triggered in many ways – by a feeling of finitude, contingency and absolute dependence; by a moment of deep appreciation of the grandeur and beauty of the natural world; by gratitude for life's goodness, or by the fellowship of suffering in time of disaster or tragedy; by reflection on 'the starry heavens above and the moral law within'; or indeed by all manner of individual promptings. When such a 'trigger' operates, there is an apperceptive switch analogous to that which occurs when, looking at a puzzle picture, we suddenly see a face where before we saw only a confusion of lines. The religious person is now experiencing-as in a new and importantly different way.

Such an apperceptive switch can produce a fairly focussed sense, momentary or prolonged, of being in the divine presence. But it can also produce a more general or diffused awareness of the world as manifesting the divine and thus as having a new meaning and value. Speaking metaphorically, the light of God's presence now shines through it. Two well known examples will suffice to illustrate this mode of experience. George Fox, the founder of Quakerism, recorded in his Journal that 'All things were new; and all the creation gave another smell unto me than before, beyond what words can utter...';[1] and Jonathan Edwards, the New England theologian, tells how 'The appearance of everything was altered; there seemed to be, as it were, a calm, sweet cast, or appearance of divine glory, in almost everything. God's excellency, his wisdom, his purity and love, seemed to appear in everything; in the sun, moon and stars; in the clouds and blue sky; in the grass, flowers and trees; in the water and all nature...'.[2]

But religious experiencing-as is not of course always theistic. Within the Mahayana tradition of Buddhism there is the startling claim that samsara, the ordinary process of human life, pervaded as it is by dukkha ('suffering', 'unsatisfactoriness'), is identical with nirvana. That is to say, experienced from the point of view of the self-concerned, grasping ego, life involves the ineradicable anxieties and fears engendered by our vulnerability to unpredictable chances and to the inevitability of decay and death. But experienced from the standpoint of the 'true' or 'original' self, found by transcending ego-centeredness and realizing one's own universal Buddha nature, life is free from these deep human anxieties. Although still finite, vulnerable and mortal, the 'awakened one' rejoices to be part of the ever-changing flow of pratitya-samutpada, the unitary interdependent life of the universe. Fr. Heinich Dumoulin quotes this description of the satori experience by a contemporary Zen priest:

> Enlightenment is an overwhelming realization which comes suddenly. Man feels himself at once free and strong, exalted and great, in the universe. The breath of the universe vibrates through him. No longer is he merely a small, selfish ego, but rather he is open and transparent, united to all in unity. Enlightenment is achieved in zazen [meditation], but it remains effective in all situations of life. Thus everything in life is meaningful, worthy of thanks, and good – even suffering, sickness and death...[3]

So far, then, we have been noting that there are religious ways, theistic and non-theistic, of experiencing both particular objects, events, and situations, and our total existence in the world, – all these being, on a religious understanding of them, modes of response to the universal presence of the Real.

As we move along the spectrum from the external to the internal, we come next to types of religious experience in which the mystic – for at this point that term begins to be commonly used – whilst being outwardly aware of the world is also inwardly aware of the Transcendent, and projects the latter awareness as a vision or audition or photism. There are numerous examples of this in the scriptures and in the post-scriptural histories of the various traditions. One example that well illustrates this mixture of the inner and the outer is the experience of the mystic, Julian of Norwich, whom I quoted earlier, as she lay apparently dying and looking at a crucifix which a priest was holding before her. Her awareness of the Real came to her in distinctively Christian terms as a sense of the divine love manifested in Jesus' death on the cross, and so fused with her perception of the wooden figure that 'suddenly I saw the red blood trickling down from under the crown, all hot, flowing freely and copiously, a living stream, just as it seemed to me that it was when the crown of thorns was thrust down upon his blessed head'.[4] Her receptiveness at this moment to the presence of the Divine, the Real, was given form (I am suggesting) by the religious ideas instilled in her by her tradition, and so created this vivid experience.

Moving yet further along the spectrum, there are forms of religious experience in which the outer world plays no part. The mystic is in a state of deep meditation or trance, unconscious of the external environment. 'To one who is entering this state', says Evelyn Underhill, 'the external world seems to get further and further away; till at last nothing but the paramount fact of his own existence remains.'[5]

In this form of religious experience, as in the others – I am suggesting – the mystic's mind is being directly affected by the divine Reality; and this impact (or, in the cybernetic sense of the term, this 'information') comes to consciousness in forms provided by the constructive imagination of the recipient, fed by the concepts and symbols of his or her tradition. If there is (as I think that in fact there is) the kind of mental impact of one human mind upon another that is called telepathy, this provides a partial analogy to the impact of the presence of the Real upon human minds. In the case of visions and auditions the telepathic analogue is that of 'crisis apparitions'. The literature of parapsychology from the days before radio communication contains numerous cases of this. Typically, a man travelling or working abroad, say in India, is unexpectedly killed in some accident. It takes two or three weeks for the news to reach his wife in England by the normal channels. But during the night immediately after the accident she has a vivid dream or a waking vision that symbolises his death. This may take a variety of forms: a dream or an apparition of her husband looking still and deathlike, or of a coffin or other symbol of death, or of receiving an announcement of his death, and so on.[6] In such a case it seems reasonable to suppose that in the moment of crisis the husband's state of mind telepathically affected that of his wife at an unconscious level and that (usually) at night, when her attention was withdrawn from the outer world, this 'information' came to her consciousness by means of the mechanism by which ordinary dreams are created. In the case of mystical experiences, however, the source of the 'information' is the universal presence of the Transcendent, the Divine, to which the mystic is exceptionally open; and this impact comes to consciousness through the basic religious concepts and the specific concrete symbols of the mystic's tradition.

What, however, of unitive mysticism? Does not this fall outside any analysis of religious experience as involving the interpretive concepts of the experiencer? For it would seem that here, without the mediation of the world or of either outer or inner visions, there is a direct experience of being absorbed into or becoming one with God, the One, the Divine. As Meister Eckhart, perhaps the greatest of the Christian unitive mystics, expressed it, 'when the divine light pours into the soul, the soul is united with God, as light blends with light'.[7] Reflecting on such experiences Eckhart says, 'If I am to know God directly, I must become completely He and He I: so that this He and this I become and are one'.[8] And so he says that 'God and I are One... I am the unmoved Mover, that moves all things'.[9] This parallels the famous Ana 'l-Haqq (I am the Real, i.e. God) of the great Muslim mystic Al-Hallaj.

From the standpoint of orthodox Christian or Muslim theology, the mystic's sense of union with God can only be subjective. For the creature remains forever ontologically distinct from and dependant for existence upon the Creator. Thus Bernard of Clairvaux, speaking of the 'unitive' state, says, 'How can there be unity where there is plurality of natures and difference of substances? The union of God and man is brought about not by confusion of natures, but by agreement of wills. Man and God, because they are not one substance or nature, cannot be called "one thing" ... but they are with strict truth called "one spirit" if they adhere to one another with the glue of love.'[10]

Thus the language of total and unqualified unity is, from this point of view, rhetorical exaggeration. In the cases of Eckhart and Al-Hallaj their respective political-religious authorities persecuted them for these statements and Al-Hallaj was even executed.

On the other hand, parallel statements by advaitic Hindu mystics are entirely acceptable within their tradition. Thus in the Crest Jewell of Discrimination (Viveka-Chudamani), attributed to Shankara, we read. 'The ego has disappeared. I have realized my identity with Brahman... My mind fell like a hailstone into that vast expanse of Brahman's ocean. Touching one drop of it, I melted away and became one with Brahman.'[11]

However this notion of a direct unitive experience of the Real, the One, the Divine, gives rise to an epistemological dilemma. An experience that is reported, and that has therefore been remembered, even if it cannot be adequately captured in words, was by definition an episode in the history of the reporter: the mystic undergoes the experience and is subsequently able to remember having had it. But in that case it seems that the mystic must never have ceased to exist as a distinct stream of consciousness. There can have been no losing of identity through becoming totally merged into the Infinite. For if the finite consciousness of the mystic had been dissolved in the Infinite, like a drop of water becoming part of the ocean, there would be no continuous thread of finite consciousness such as is required for an individual memory of the experience. We therefore seem driven to conclude either that the remembered 'unitive experience' was not truly unitive or that, if the mystic was indeed absorbed into a truly unitive state, this cannot have been properly speaking an experience undergone by the mystic, since he or she could then have no memory of it.

It therefore seems to me that the remembered experience, which is subsequently spoken of as unitary, must have been the experience of a continuing finite consciousness, though one in which the entire field of consciousness became filled, in the case of a theistic mystic, by the presence of God, and in the case of the advaitic mystic, by the universal reality of Brahman. For it would seem that when a human self has attained to a sufficient transcendence of the ego and its concerns it may become so open to the Divine, the Real, the Ultimate as to undergo moments in which the consciousness is totally filled with the transcendent Reality. But the fact that this 'unitive' experience takes both theistic and non-theistic forms (and sub-varieties of each) suggests that the finite consciousness of the mystic has carried with it into this moment the basic conceptual structure whereby it experiences the Real, in some cases as a personal, and in other cases as a non-personal, reality.

I am suggesting, then, that religious experience in all its forms is a mode of consciousness that occurs when someone is freely (though not necessarily by conscious volition) open and responsive to the universal presence of the transcendent Reality. The impact of this presence comes to consciousness as a mode of experience whose specific forms are provided by the experiencer's religious concepts and symbols. This of course would explain how it can be that Christians have distinctively Christian forms of religious experience, Muslims distinctively Islamic forms, Hindus distinctively Hindu forms, and so on.

But let us now look at the same range of phenomena from a naturalistic point of view, which denies that there is any transcendent Reality to evoke these forms of experience. A naturalistic philosophy must hold that religious experience is internally generated. Instead of religious concepts being used to give form to our awareness of a transcendent impact upon us, they are being used to shape the products of our own creative imaginations, the (unconscious) motive behind their production being primarily reassurance in the face of our inescapable finitude, vulnerability and mortality.

Since the same reported experiences occur on either a religious or a naturalistic view, wherein lies the difference between situations in which the one or the other of these views is true? What is the difference between there being and there not being a transcendent divine Reality which is a factor, in collaboration with our human conceptual systems, in the genesis of religious experience? The answer must lie, not in the nature of the experiences themselves as describable phenomena, but in the implications of their rival construals for the larger structure of the universe considered as entailing the possibility of further experiences. If a religious account of the universe is true, the universe is so structured as to give rise to further human experiences in the future which will not occur if a naturalistic account is true. For each of the great traditions teaches that our present earthly life is only a small segment of our total existence. Each tradition speaks of an ultimate eschatological state, variously pictured as heaven or hell, paradise, oneness with the infinite consciousness of Brahman, or as a nirvanic state beyond individual ego-existence. Whilst there are these, and yet other, pictures and symbols, strong strands of each tradition regard the ultimate state as lying beyond the range of our earthly concepts. However each also speaks in more concrete terms of the pareschaton (the phase or phases between this life and that ultimate state) in which we continue to exist as individual persons, being either resurrected or reborn in this or other worlds, or going through some kind of purgatorial experience or 'continued sanctification after death'. We do not have to choose between these different pareschatologies in order to see that if any one of them is correct a purely naturalistic understanding of the universe must be incorrect. Indeed this is still the case if none of them is correct in its specific expectations and yet all are correct in their common affirmation of continued personal existence beyond the present life. It could further be the case that as we proceed into whatever unexpected form our continued existence takes we find that a religious, as distinguished from a naturalistic, understanding of the universe is progressively confirmed whilst the particular conceptualization of that religious understanding is itself being progressively revised or transcended. It is at any rate clear that the wider sets of beliefs and expectations that cohere respectively with a religious and with a naturalistic construal of religious awareness are very different and cannot both be compatible with the full range of human experience.

Needless to say, I have only touched in these last remarks upon a very large field for speculation. There can be a variety of possible complications of this picture, modifying naturalism in the religious direction. But I am only concerned here to make the general point that the rival understandings of religious experience are integral to radically different conceptions of the larger character of the universe, involving different expectations concerning the context and the content of future human experience.

The claim that religious experiences connect with expectations of some kind of life after death does nothing, of course, to establish now the truth of this larger conception of the universe. It does however show that, whilst religious and naturalistic believers grant that the various reported forms of religious experience occur, and are thus far in agreement, their different understandings of those experiences are such that if the religious understanding turns out to be correct the naturalistic understanding will thereby have been shown to be false.

In the meantime the two construals of religious experience – as cognitive in humanly varying ways of a transcendent Reality, and as purely a product of the human imagination – connect with different forms of life, different ways of inhabiting the world. These often have considerable overlap on the physical and ethical levels of meaning. But they differ basically in their understanding of the over-all character of the universe and hence of what is going on in and through the entire phenomenon of human existence.

Each of these global interpretations constitutes an act of faith: the religious and the naturalistic believer must both be said to 'walk by faith'. The religious believer is living in terms of a meaning which may prove to be substantially true or may be an illusion; and the naturalistic believer is living in terms of a meaning which may be true or may prove to be an illusion. And we have to say that both are at present entitled to believe as they do, and that each is taking a cognitive risk is no doing.

NOTES

1 The Journal of George Fox, p. 17.
2 Quoted by William James (1960), p. 248.
3 Heinrich Dumoulin (1963), p. 275.
4 Showings, short text, chap. 3.
5 Evelyn Underhill (1955), p. 318.
6 For actual accounts see, e.g., F. W. H. Myers (1975).
7 Sermon 14, Blakney (1941), p. 163.
8 Sermon 99, quoted by Evelyn Underhill (1955), p. 420.
9 Sermon 28, Blakney (1941), p. 232.
10 Bernard of Clairvaux, Sermons on the Canticles, Sermon 71. See also St. John of the Cross, Ascent of Mount Carmel, Book II, Chap. 5, and John Tauler's 'First Sermon for the Second Sunday of the Epiphany' in A. Poulain (1950), chap. 18.
11 Shankara (1978), p. 113.

REFERENCES

Bernard of Clairvaux: *Sermons on the Canticles, Sermon 71.*

Dumoulin, Heinrich: 1963, *A History of Zen Buddhism* [1959], trans. Paul Peachey, Random House, New York.

James, William: 1960, *Varieties of Religious Experience* [1902], Collins, London/Mentor Books, New York.

Julian of Norwich: 1978, *Showings*, trans. Edmund Colledge and James Walsh, Paulist Press, New York.

The Journal of George Fox 1994, J. M. Dent, London/E. P. Dutton, New York 1924.

Meister Eckhart: *Sermon 14, Sermon 28*, trans. Raymond B. Blakney 1941, Harper & Row, New York.

Myers, F. W. H.: 1975, *Human Personality and its Survival of Bodily Death* [1903], Arno Press, New York.

Poulain, A.: 1950, *The Graces of Interior Prayer, Part II*, B. Herder, St. Louis.

Shankara: 1978, *Crest Jewell of Discrimination*, trans. Swami Prabhvananda and Christopher Isherwood, Vedanta Press, Los Angeles.

St. John of the Cross: *Ascent of Mount Carmel, Book II.*

Tauler, John: 'First Sermon for the Second Sunday of the Epiphany' in A. Poulain (1950), chap. 18.

Underhill, Evelyn: 1955, *Mysticism* [1911], New American Library, New York.

DISCUSSION

WEINGARTNER: Let me ask you to elaborate the point of the cognitive risk a little bit more, because this would also be interesting in respect to the similarities between scientific and religious belief I've spoken about.

HICK: What I am pointing out is that the universe is ambiguous in respect of its religious or non-religious significance. That is to say, one can give a complete and coherent account of it in religious terms, but one can also give an equally complete and equally comprehensive account of it in naturalistic terms. The naturalistic account will include religious experience and religious history, the religious account will include the naturalistic response. Each can, so to speak, swallow the other. And from our present standpoint within the universe we have no way of establishing which is correct. But, second, it also seems to me, that the people who experience religiously are fully entitled rationally to form their beliefs and base their lives on the religious mode of experiencing. But it is equally rational, i.e. equally sane, for the person who does not experience religiously at all, not to form religious beliefs. Thus each of them is taking a cognitive risk, of which, if they are reflective, they will be aware, namely the risk of being fundamentally mistaken. And then the third thing I want to say, just to complete the circle, is that the universe does in fact have either a religious or a naturalistic character. However, if it has a naturalistic character, we shall never definitively know this. It will remain ambiguous to us until the moment of our death. On the other hand, if it has religious character, that it has this character will become evident to us. Thus we are not simply dealing with that Hare called 'blicks', i.e. different optional ways of seeing the universe. What we are dealing with is ways of seeing, one of which is right and the other of which is wrong, though at the moment we have no way of establishing which. Thus in basing ourselves on the way we experience it, whether naturalistically or religiously, we are taking a risk which can't be avoided. With regard to this point, William James argued very effectively, I think.

WEINGARTNER: Do you mean that for a religious person this risk is also extended to the creed? In my view, as a consequence of his belief, the religious believer also believes that in basic propositions of the creed, he cannot be wrong. Having raised the problem of cognitive risk you refer more to the general religious attitude towards the whole interpretation of the universe which is of course a different thing than that to believe in a certain creed.

HICK: I would say, that the risk involved in believing a specific creed is much greater, – that is, the possibility of being wrong is much greater. In the case of the basic religious perception there is a 50:50 chance of being right, but when you come down to a very specific creed (bearing in mind that there are hundreds of these creeds), the possibility of being wrong is multiplied. I mean it does not seem to me very likely that among the many different religious belief-systems someone is absolutely right and all the others wrong. But it could be the case, of course.

WEINGARTNER: I understand your point, but my question was in addition, whether the believer himself has this view that the probability decreases so much or whether, in respect to his creed, he is sure that it cannot be wrong.

HICK: If we are speaking of psychology, there certainly are a lot of people who are absolutely sure that their creed is totally true. But I feel myself that if one is reflective, one will also be conscious that one may be mistaken.

BOCHENSKI: If I understood you correctly, the term "religious" appears in your construction on two levels. The first one presupposes that there is a sort of religious experience which will be accepted as a fact, even by naturalists. And then there are two possible interpretations of the world including that religious experience: the naturalistic way and the religious one. Thus you assert the existence of religious experience as a fact which cannot be denied.

HICK: Yes. And it seems to me that the basic principle of cognitive sanity, or rationality, is to accept that things are as they seem to be except, and this of course is extremely important, when we have some positive reason to think that things are not as they seem to be. The same principle will apply not only to sense experience but to religious experience as well. Therefore, it will be insane or irrational for – let us take a very extreme case – a great saint, who lives all the time in the presence of God, not to believe in the reality of God, just as it would be insane for us not to believe in the reality of these tables.

LEHRER: I have two points. The first point I want to take up concerns the cognitive risk. You suggested that the cognitive risk involved with theism is a rational one even if the probability is only 50:50.

HICK: Yes.

LEHRER: I have some doubts about that. Take the hypothesis for instance, that the number of trees in Salzburg is even. Well, if you are interested in having a true view about whether the number of trees in Salzburg is even or not, you might say, 'Well, you may as well go ahead and accept it', I mean, nothing ventured, nothing gained. But that does not somehow meet with our prior conception of rationality. And the reason I think it doesn't is that the cognitive risk of error is too high.

HICK: Yes, right. But here I would appeal to the reasoning of William James, who strikes me as an enormously sensible mind. In his famous essay "The Will to Believe" he points out the difference between the religious case and this other type of case of the number of trees in Salzburg. In the former we are faced with the forced option, that is, we have to proceed on the basis that the universe is religiously structured or not. So it would not be possible to say, 'I am not going to decide', because in not deciding we are deciding. We are in fact choosing the naturalistic option if we do not choose the religious one. Second, it makes a big difference in that, that if the world is religiously structured, we are very much better off if we proceed on that basis. It does not matter to us whether the number of trees in Salzburg is odd or even. But it does matter in a very total and profound way whether we are going to live on the basis that the world in which we live has an ultimately good meaning or it is just a sort of jungle in which we hope for the best.

LEHRER: There are two things that occur to me. One is, and this may just be verbal, but it sounds like one is taking more of a practical risk than a cognitive one. I mean, cognitive risk I would think of as being a risk that's only concerned with the objective of obtaining truth and avoiding error. And you have introduced another component. It looks like practical decision making rather than epistemic decision-making.

HICK: It seems to me to be both. I quite agree, it is not purely cognitive in the narrow sense. It is a question of seeing the world in a particular way, as having a certain meaning, and living in terms of that meaning. But you are absolutely right, of course, that it is also practical.

LEHRER: Here is the second question. It seems to me, that if on the basis of his experience the theist conceives the religious picture to be more probable than the naturalistic picture, then in fact the probability assignment represents a prior probability assignment, i.e. that assigns higher probability than a half to the religious hypothesis. What I'd like to argue is that there is a background system that is presupposed, and you have to choose between these systems. I think it is reflected rather nicely in probability theory. Thus, if you are a religious person, you think that the conditional probability of the religious hypothesis is higher. But if you are a naturalist, you think that the naturalistic hypothesis has a higher probability. I think what you can demonstrate, and it is perhaps interesting and sympathetic to the paper that you are presenting, is that conditional probabilities are to be dependent on the basis of a prior probability. So, it's to me a theorem of the calculus of probability where if somebody assigns a higher probability to the religious hypothesis on the basis of their experience, that's, because they assign a higher probability to the religious hypothesis as a prior probability, as a sort of basic or fundamental assumption, and *vice versa* with respect to the naturalistic one.

HICK: And you mean, they do this on the ground of, or lack of, religious experience?

LEHRER: To defend the view that the religious picture is true on the basis of your experience, you really have to assign a higher probability to the religious hypothesis to begin with. And the naturalist is, of course, in the same boat.

HICK: Well, although this is certainly sympathetic in intention, I am inclined to feel that it would lead us into trouble, because it leads us back to the rationality in the sense of a good inference from data. But, it seems to me, the difference between the religious and the non-religious person is in their different data, i.e. presence or absence of religious experience. Each is being rational given the data they have. All I want to claim is that the religious person is being perfectly rational in going on his data, granting that the non-religious person is equally rational in going on his different data.

LEHRER: Aha, I think, the point is sympathetic to that, because what you make of the data is going to depend on your prior probabilities. They may be in themselves influenced by experience, in the sense that probabilities are not something that is determined a priori by logic or something like that. They have to be taken as basic assumptions between some other worldly experience, but the point I am making is that if the naturalist claims that her experiences render naturalism probable, then this claim suggests her assigning a higher prior probability to the naturalism.

HICK: I see your point. The question, however, is whether it fits the facts, I mean, whether it is the case that the religious person, prior to any kind of participation in a stream of religious experience, gives a higher initial probability to theism. Not necessarily, I think. That may have been so in the past, but probably not so today in our society. What does sometimes happen is that people, starting out as naturalists, have undergone a form of religious experience without any prior view that the theistic view is initially more probable. Of course one can also talk in the way you have been suggesting about sense perception, yet it seems to be very contrived. When I see a table, I don't think that there is a higher probability that there is a table than that there is not. I just see the table and live on the basis that there is a table there. To translate that into terms of probability does not seem to me necessary or natural. Would you apply all this to sense perception?

LEHRER: I think it is natural in the following way: if you have certain kinds of experiences, then you fall back on an assumption which may be built in biologically as far as I know. In certain very simple perceptional cases the situation is such that when I have certain sorts of experiences, then I form some kind of belief.

HICK: Yes.

LEHRER: But that can be reflected in my assigning probability to that sort of belief being true I have a kind of experience.

HICK: That is something you would do if you are interested in probability theory, isn't it? But when the stone-age person sees the mammoth rushing towards him, he doesn't consider probabilities but he just thinks how to survive. And I think all of us normally behave this way too.

BOCHENSKI: I think, you said that the probability of accepting a religious system as opposed to basic religion as accepting a religious system as true, is rather low.

HICK: Yes it seems to me that in the philosophy of religion we have to take account of the plurality of religious systems. Otherwise, what we are doing is the theology of our own system. But now a big problem arises from the fact that there are different sets of beliefs, different sets of truth-claims. From a naturalistic point of view one can say that all of them are wrong. But what are we to make of this plurality from a religious point of view? So far as we can tell each of the great world religions is more or less on a par in their beneficial effects upon human beings, in changing human beings from self-centredness to a new orientation centred in the divine reality as they perceive it. So it is not an easy option to say that belief-system A is right and all the others wrong. The sort of line that opens up is one in which one postulates an ultimate divine reality that in itself is beyond the scope of our human concepts. This is something, that is all the great traditions ultimately say. But that reality is humanly perceived and humanly conceived in very different ways. The belief systems, then, spell out particular ways of conceiving the divine. But the situation may not be that only one of them is right. They may all be right as spellings out of a valid human perception of the divine. Perhaps we should abandon the assumption that there is only one valid human perception of the divine. There may be many. And when I said, the probability of anyone's system being right is reduced because there are so many systems, I meant the probability of that system being the whole and sole truth.

BOCHENSKI: Yes, but there are contradictions between the systems.

HICK: Well, and if they are all about the ultimate in itself, then yes, they contradict each other: the ultimate can't be both personal and non-personal, or both be Trinity and unity. But on the other hand, if what they are each about is a particular perception of the divine, then the fact that there is one perception of the divine does not necessarily contradict the fact that there is another perception of the divine which is different but which may be equally a channel between the human and the divine.

BOCHENSKI: What would you say about conception, which may appear as a reasonable one, that while basic religion may be true with some qualification, namely analogy in the profane meaning of the term. The concept of analogy has been elaborated with use of mathematical-logical means both by myself and by Prof. Weingartner. When you say that God is our father, you do not assert that physiological and sociological relations holding between a human father and his child, do also hold of God; – only some relations isomorphic to them. But, with that restriction, I think, that the creed of a basic religion taken literally may be true. The religious systems would be, on the contrary, only metaphorical expressions of the basic faith. As an example, take the old-Egyptian myth of Osiris, who is said to have been killed and cut in fourteen pieces after he had begot a son; his wife Isis collected them and set them together. Such a myth can certainly not be literally true. It can be, however, metaphorically true. Then it means that man can somewhat overcome death, a proposition which may be literally true. What I am trying to suggest is, that whereas the creed of a basic religion may be true, in its literal (if analogical) meaning, the religious systems can be only its metaphorical expressions.

HICK: That's right. However, I realize that we have different conceptions of basic religion. I think, for you basic religion is essentially theistic: is that right?

BOCHENSKI: No.

HICK: Well, you spoke of God and "his", in personal terms.

BOCHENSKI: I think that God can be conceived analogically as a person, and thus spoken of in personal, analogical terms.

HICK: Right, yes. I would say that our Christian understanding of God, and likewise the Islamic one, and also the Buddhist concept, which is not a concept of God at all, are all symbolic structures, all of them.

BOCHENSKI: Symbolic meaning either in the way as let's say, God in the Egyptian religion or symbolic meaning in the sense of analogy? These are two quite different symbolic meanings.

HICK: Well, actually let me withdraw the word "symbolic". I used it because you were using it, but I'd rather put it this way, that any set of human concepts applied to the ultimate is, if you like, mythic or metaphorical.

BOCHENSKI: But I think that it is symbolic in a much stronger sense of "symbolic", namely meaning here "analogical". In my above example the proposition "man can overcome death" which is the metaphorical meaning of the myth of Osiris, must not have again just a metaphorical meaning.

HICK: Well, I think, that there was, for example, Jesus of Nazareth who was a historical person, and I think that something happened after his death that has given rise to the stories of resurrection, but it is impossible today to know exactly what it was that happened.

BOCHENSKI: That's why I am not so envious to ask you about your personal beliefs. The question was, if you don't think the natural solution of the problem of plurality of religious systems consists in saying that whereas the creed of the basic religion may be true in some analogical way, it is not so about the systems, where only metaphorical meaning can be admitted.

HICK: You are right, yes. But would you not also be willing to say that these different symbol systems or mythological systems serve a very important and positive function; they enable us to think and live in relation to the divine. And the important question about them is not, whether they are literally true or not, but whether they function effectively.

BOCHENSKI: That view does not deny the utility of religious systems. The important point is however, that it is not identical with the modernistic position, according to which they are symbols of moral rules only. Would you admit that what I suggested is a possible explanation of the plurality of religious systems and their capacity of prediction?

HICK: Yes.

BOCHENSKI: You say yes. Thank you.

HICK: Well, we are still thinking to some extent in different ways, though in contrast with the way many people think we are thinking in the same way.

WEINGARTNER: One important point was, that you said, the difference between a religious person and a non-religious person is that they have just different data.

HICK: Yes.

WEINGARTNER: But if they have different data, they would also have different conclusions from these data. Also they might have different kinds of explanations or hypotheses interpreting these data, whereas this is not always necessary, it depends on how different the data are, because from premises you can always derive different conclusions and if the hypotheses are wide enough, they could perhaps "explain" (in a loose way) very different data.

HICK: Yes.

WEINGARTNER: So from different data we could not conclude that the interpreting hypotheses of these ultimate explanations which we have in religion but also in 'Weltanschauung' have to be also different.

HICK: Yes, if it were a question, as a whole of course tradition in philosophy religion takes it to be, – of interpreting in the sense of finding the best explanation for, the data, then what you are saying is absolutely correct. But I don't think that being religious, or becoming religious, is a matter of drawing certain conclusions and thinking that this conclusion is more, or less, probable. That's what philosophers do, but this is not what you find in descriptions of any of the religious traditions. What you find I think, is mode of experiencing. Take a very strong case: Jesus was all the time intensely conscious of being in the presence of God, to such an extent that it would have been insane for him to deny the reality of God and was totally sane for him to affirm the reality of God. But it was not a matter of his thinking it more probable that there is a God. That whole way speaking, I think, takes us out of religion into something else, namely philosophy, or perhaps into theology. How does that strike you?

WEINGARTNER: I think, I have the impression that this depends on this very question, whether religious people have more or less of that kind of what you call religious experience. I mean, if I think of great theologians in the Middle Ages, then you would probably say that they are not the religious persons in the sense you have in mind.

HICK: Yes, in some cases.

WEINGARTNER: They are in addition theologians and in addition philosophers and so on. So, at least, if I think of those people, then I would like to say, that they try hard to have these higher-order hypotheses or global explanations and a whole system which can explain in some objective way these kinds of religious data or phenomena.

HICK: Yes.

WEINGARTNER: So it would be probably different from one category of religious people to another.

HICK: If we use the word in a very broad sense, in which you look at a map of the world and say that everyone living in North America is a Christian, and everyone living in Saudi-Arabia is a Muslim, and so on, then you are going to have to distinguish between first-hand and second-hand religion, and to say that many people have a second-hand religion, i.e. not based on their own religious experience but instilled into them by their society. And most religion perhaps is of that character. But there would be no second-order religion, second-hand religion, if there were not also first-hand religion. Christianity, for example, would have died out if it was simply second-hand religion. It all lives from the first-hand religion of the saints.

WEINGARTNER: But the term "saint" acquires a special connotation in your account. I mean, if you take Franciskus, he would develop a kind of the first-order religion, according to you, but what's about Albert the Great for instance, who was a natural scientist ?

HICK: Yes, I do not mean by "saint" someone whom the church declares to be a saint. Rather, I myself think of several people, two or three of whom are Christians, one of whom is a Sikh, several of whom are Buddhists, one or two of whom are Hindus, one of whom is Jewish, two of them are Muslims, who may be called saints in the sense that they are spiritually enormously superior to ourselves, people we would love to be like. This is something extremely hard to codify, but very evident in the concrete case. There are people who are not thinking of themselves and their own interests at all, but – using metaphors now – they are transparent to God, or to a higher reality of some kind, that affects their whole lives. Now it seems to me that if the religious traditions did not produce such people, they would have become simply sociological phenomena long ago, and indeed to great extent they are sociological phenomena.

LIST OF CONTRIBUTORS

Professor JOSEPH M. BOCHENSKI, 2, Square des Places, CH-1700 Fribourg.

Joseph Maria Bocheński was born in Czuszow, Polen, in 1902. He studied jurisprudence at the University of Lemberg (1920-22); economics at the University of Posen (1922-26); philosophy at the University of Fribourg (1928-30), where he acquired a Ph.D.; and theology at the Angelicum in Rome (1930-33), where he acquired a D.D. He was Professor at the Angelicum in Rome (1933-40), at the University of Krakau (1938), and at the University of Fribourg (1945ff.), where he was Dean of the Philosophical Faculty (1950-52) and President (1964-66). He was Director of the *Osteuropäisches Institut* (1957) and Director of the *Ost-Kolleg* in Cologne (1955-65). He was Visiting Professor at the Universities of Notre Dame (Flint Professor), Kansas, Pittsburgh, Alberta, Salzburg, and South Africa. He has received honorary doctorates from the University of Notre Dame (jurisprudence), from the University of Mexico (philosophy), and from the University of S. Cuore in Milan (mathematics). His interests include all core disciplines of philosophy and theology. His most recent work concentrates (again) on Thomas Aquinas. He is the author of numerous philosophical articles and books. Perhaps his most important books are: *Europäische Philosophie der Gegenwart*, 1947; *Formale Logik*, 1956; *Logik der Religion*, 1964.

Professor JOHN HICK, Institute for Advanced Research in the Humanities, University of Birmingham, 144 Oak Tree Lane, Selly Oak, Birmingham B29 6HU.

John Hick was born in Scarborough, England, in 1922. He studied philosophy at the Edinburgh University (M.A., and D.Litt.) and Oxford University (D.Phil.). He has taught the philosophy of religion at Cornell University, Princeton Theological Seminary, Cambridge University, Birmingham University, and the Claremont Graduate School, California, of which he is Danforth Professor Emeritus. He has written on a number of aspects of the philosophy of religion in *Faith and Knowledge*, 1957, 1966; *Philosophy of Religion*, 1963, 1973, 1983, 1990; *Evil and the God of Love*, 1966, 1977; *God and the Universe of Faiths*, 1973; *Death and Eternal Life*, 1976;

Problems of Religious Pluralism, 1985; *An Interpretation of Religion*, 1989, based on the Gifford Lectures for 1986-87; *Disputed Questions*, 1993; etc. He has also written on theological topics in, for example, *The Metaphor of God Incarnate*, 1993. His main research interests are in the epistemology of religion and in the problems of religious pluralism.

Professor ELENA KLEVAKINA-ULJANOV, International Research Center Salzburg, Mönchsberg 2a, A-5020 Salzburg.

Elena Klevakina-Uljanov was born in Novosibirsk in 1956. From 1973 on she studied at the Pedagogical Institute, Faculty of Foreign Languages, and achieved her diploma with distinction in English and German in 1978. From 1978 to 1981 she was a Lecturer in English at the Medical Institute of the University of Novosibirsk. In 1981 she began her postgraduate studies in philosophy, was a Lecturer in philosophy at the Medical Institute of the University of Novosibirsk from 1983 to 1984, and obtained a master-degree in philosophy in 1984. In 1985 she became a Senior-Teacher in philosophy at the Department of Philosophy of the Pedagogical Institute of the University of Novosibirsk. From 1985 to 1990 she was Dean of the Faculty of Foreign Languages of the University of Novosibirsk, and has been a Docent (Assistant Professor) of Philosophy at the University of Novosibirsk since 1986. During the winter-term 1989/90 and the summer-terms 1990 and 1991 she was Visiting Professor at the Department of Philosophy of the University of Salzburg. Since winter-term 1991/92 she has been Visiting Professor at the Department of Philosophy at the University of Vienna. Her main interests are in epistemology and the history of Russian philosophy. She has a number of publications in Russian and English – for instance: 'Epistemic Concepts and the Problem of Rationality', in Paul Weingartner and Gerhard Schurz (eds.), *Philosophy of the Natural Sciences. Proceedings of the 13th International Wittgenstein-Symposium*, Vienna, Hölder-Pichler-Tempsky, 1989 and 'On the Problem of Reducing Value-Components in Epistemology', in E. Agazzi (ed.), *The Problem of Reductionism in Science*, Dordrecht, Kluwer, 1991.

Professor FRANZ VON KUTSCHERA, Department of Philosophy, University of Regensburg, Universitätsstraße 31, D-93040 Regensburg.

Franz von Kutschera was born in Hannover, Germany, in 1932. He studied physics and philosophy in Munich, obtained his Ph.D. there in 1960 and his habilitation degree in 1963. Since 1968 he has been Full Professor of Philosophy at the University of Regensburg, Bavaria. His main fields of interest are philosophical logic, epistemology, philosophy of language and ethics. Besides numerous papers, he has published several books, for instance: *Wissenschaftstheorie* (2 volumes) München, Fink, 1972; *Einführung in die Logik der Normen, Werte und Entscheidungen*, Freiburg, Alber, 1973; *Sprachphilosophie*, München, Fink, 2nd ed. 1975; *Einführung in die intensionale Semantik*, Berlin, de Gruyter, 1976; *Grundfragen der Erkenntnistheorie*, Berlin, de Gruyter, 1982; *Grundlagen der Ethik*, Berlin, de Gruyter, 1982; *Gottlob Frege*, Berlin, de Gruyter, 1989; and *Vernunft und Glaube*, Berlin, de Gruyter, 1990.

Professor KEITH LEHRER, Department of Philosophy, University of Arizona, Tucson, 85721.

Keith Lehrer was born in Minneapolis, Minnesota in 1936. He studied philosophy at the University of Minnesota and Brown University where he received his doctorate in 1960. He is currently Regents Professor of Philosophy at the University of Arizona, Honorary Professor at Karl-Franzens-Universität Graz, and Chairman of the Board of Officers of the American Philosophical Association. His main areas of research are freedom, knowledge, rational consensus and the philosophy of Thomas Reid. His most important books are *Knowledge*, Oxford, Clarendon Press, 1974; *Rational Consensus in Science and Society* (with Carl Wagner), Dordrecht, D. Reidel Publishing Company, 1981; *Thomas Reid*, London, Routledge, 1989; *Theory of Knowledge*, Boulder and London, Westview Press and Routledge, 1990 and *Metamind*, Oxford, Clarendon Press, 1990. His work is the subject of two books: *Keith Lehrer*, edited by R. Bogdan, Dordrecht, D. Reidel Publishing Company, 1981 and *Metamind, Knowledge and Coherence: Essays on the Philosophy of Keith Lehrer*, edited by J. Brandl (et al.), Amsterdam, Editions Rodopi, 1991.

Professor GERHARD SCHURZ, Department of Philosophy, University of Salzburg, Franziskanergasse 1, A-5020 Salzburg.

Gerhard Schurz was born in Graz, Austria, in 1956. He studied first chemistry and then philosophy at the University of Graz. Diploma in theoretical chemistry in 1980. Ph.D. in philosophy in 1983. Thesis: *Wissenschaftliche Erklärung. Ansätze zu einer logisch-pragmatischen Wissenschaftstheorie*, Graz, 1983. Habilitation on *Relevant Deduction in Science and Ethics. With a Case-Study of the Is-Ought Problem* in 1989 (publication in progress). His main interests are in philosophy of science, philosophical logic, cognitive science and artificial intelligence, epistemology, philosophy of natural sciences and history and ethics of natural sciences. He has published one book and more than 50 papers.

Professor PAUL WEINGARTNER, Department of Philosophy, University of Salzburg, Franziskanergasse 1, A-5020 Salzburg.

Paul Weingartner was born Innsbruck, Austria in 1931. He studied philosophy, physics, mathematics and pedagogics at the University of Innsbruck. Ph.D. 1961, major: philosophy, minor: physics. Postdoctoral studies in London (with K. R. Popper) and in Munich (with Stegmüller). Habilitation on *Basic Questions on the Problem of Truth* (University of Graz, 1965). His research areas are philosophy of science, logic, epistemology and investigations on traditional problems of philosophy with the help of modern tools of logics and mathematics. He is author of two books on philosophy of science and of more than 100 research articles. He is chairman of the Institut für Wissenschaftstheorie (Institute for the Philosophy of Science) at the International Research Center Salzburg (Mönchsberg 2a, A-5020 Salzburg).

PHILOSOPHICAL STUDIES SERIES

Founded by Wilfrid S. Sellars and Keith Lehrer

1. JAY F. ROSENBERG, *Linguistic Representation,* 1974.
2. WILFRID SELLARS, *Essays in Philosophy and Its History,* 1974.
3. DICKINSON S. MILLER, *Philosophical Analysis and Human Welfare.* Selected Essays and Chapters from Six Decades. Edited with an Introduction by Lloyd D. Easton, 1975.
4. KEITH LEHRER (ed.), *Analysis and Metaphysics.* Essays in Honor of R. M. Chisholm. 1975.
5. CARL GINET, *Knowledge, Perception, and Memory,* 1975.
6. PETER H. HARE and EDWARD H. MADDEN, *Causing, Perceiving and Believing.* An Examination of the Philosophy of C. J. Ducasse, 1975.
7. HECTOR-NERI CASTAÑEDA, *Thinking and Doing.* The Philosophical Foundations of Institutions, 1975.
8. JOHN L. POLLOCK, *Subjunctive Reasoning,* 1976.
9. BRUCE AUNE, *Reason and Action,* 1977.
10. GEORGE SCHLESINGER, *Religion and Scientific Method,* 1977.
11. YIRMIAHU YOVEL (ed.), *Philosophy of History and Action.* Papers presented at the first Jerusalem Philosophical Encounter, December 1974, 1978.
12. JOSEPH C. PITT, *The Philosophy of Wilfrid Sellars: Queries and Extensions,* 1978.
13. ALVIN I. GOLDMAN and JAEGWON KIM, *Values and Morals.* Essays in Honor of William Frankena, Charles Stevenson, and Richard Brandt, 1978.
14. MICHAEL J. LOUX, *Substance and Attribute.* A Study in Ontology, 1978.
15. ERNEST SOSA (ed.), *The Philosophy of Nicholas Rescher: Discussion and Replies,* 1979.
16. JEFFRIE G. MURPHY, *Retribution, Justice, and Therapy.* Essays in the Philosophy of Law, 1979.
17. GEORGE S. PAPPAS, *Justification and Knowledge: New Studies in Epistemology,* 1979.
18. JAMES W. CORNMAN, *Skepticism, Justification, and Explanation,* 1980.
19. PETER VAN INWAGEN, *Time and Cause.* Essays presented to Richard Taylor, 1980.
20. DONALD NUTE, *Topics in Conditional Logic,* 1980.
21. RISTO HILPINEN (ed.), *Rationality in Science,* 1980.
22. GEORGES DICKER, *Perceptual Knowledge,* 1980.

23. JAY F. ROSENBERG, *One World and Our Knowledge of It*, 1980.
24. KEITH LEHRER and CARL WAGNER, *Rational Consensus in Science and Society*, 1981.
25. DAVID O'CONNOR, *The Metaphysics of G. E. Moore*, 1982.
26. JOHN D. HODSON, *The Ethics of Legal Coercion*, 1983.
27. ROBERT J. RICHMAN, *God, Free Will, and Morality*, 1983.
28. TERENCE PENELHUM, *God and Skepticism*, 1983.
29. JAMES BOGEN and JAMES E. McGUIRE (eds.), *How Things Are, Studies in Predication and the History of Philosophy of Science*, 1985.
30. CLEMENT DORE, *Theism*, 1984.
31. THOMAS L. CARSON, *The Status of Morality*, 1984.
32. MICHAEL J. WHITE, *Agency and Integrality*, 1985.
33. DONALD F. GUSTAFSON, *Intention and Agency*, 1986.
34. PAUL K. MOSER, *Empirical Justification*, 1985.
35. FRED FELDMAN, *Doing the Best We Can*, 1986.
36. G. W. FITCH, *Naming and Believing*, 1987.
37. TERRY PENNER, *The Ascent from Nominalism*. Some Existence Arguments in Plato's Middle Dialogues, 1987.
38. ROBERT G. MEYERS, *The Likelihood of Knowledge*, 1988.
39. DAVID F. AUSTIN, *Philosophical Analysis*. A Defense by Example, 1988.
40. STUART SILVERS, *Rerepresentation*. Essays in the Philosophy of Mental Rerepresentation, 1988.
41. MICHAEL P. LEVINE, *Hume and the Problem of Miracles. A Solution*, 1979.
42. MELVIN DALGARNO and ERIC MATTHEWS, *The Philosophy of Thomas Reid*, 1989.
43. KENNETH R. WESTPHAL, *Hegel's Epistemological Realism*. A Study of the Aim and Method of Hegel's *Phenomenology of Spirit*, 1989.
44. JOHN W. BENDER, *The Current State of the Coherence Theory*. Critical Essays on the Epistemic Theories of Keith Lehrer and Laurence Bonjour, with Replies, 1989.
45. ROGER D. GALLIE, *Thomas Reid and 'The Way of Ideas'*, 1989.
46. J-C. SMITH (ed.), *Historical Foundations of Cognitive Science*, 1990.
47. JOHN HEIL (ed.), *Cause, Mind, and Reality*. Essays Honoring C. B. Martin, 1990.
48. MICHAEL D. ROTH and GLENN ROSS (eds.), *Doubting*. Contemporary Perspectives on Skepticism, 1990.
49. ROD BERTOLET, *What is Said*. A Theory of Indirect Speech Reports, 1990
50. BRUCE RUSSELL (ed.), *Freedom, Rights and Pornography*. A Collection of Papers by Fred R. Berger, 1990
51. KEVIN MULLIGAN (ed.), *Language, Truth and Ontology*, 1992
52. JESÚS EZQUERRO and JESÚS M. LARRAZABAL (eds.), *Cognition, Semantics and Philosophy*. Proceedings of the First International Colloquium on Cognitive Science, 1992

53. O.H. GREEN, *The Emotions*. A Philosophical Theory, 1992
54. JEFFRIE G. MURPHY, *Retribution Reconsidered.* More Essays in the Philosophy of Law, 1992
55. PHILLIP MONTAGUE, *In the Interests of Others.* An Essay in Moral Philosophy, 1992
56. J.- P. DUBUCS (ed.), *Philosophy of Probability.* 1993
57. G.S. ROSENKRANTZ, *Haecceity:* An Ontological Essay. 1993
58. C. LANDESMAN, *The Eye and the Mind.* Reflections on Perception and the Problem of Knowledge. 1993
59. P. WEINGARTNER (ed.), *Scientific and Religious Belief. 1994*

FSC
www.fsc.org
MIX
Papier | Fördert
gute Waldnutzung
FSC® C083411

Kolibri 360 GmbH
Ferdinand-Jühlke-Straße 7
99095 Erfurt, Deutschland
produktsicherheit@kolibri360.de